SpringerBriefs in Earth System Sciences

Series Editors

Gerrit Lohmann, Universität Bremen, Bremen, Germany

Justus Notholt, Institute of Environmental Physics, University of Bremen, Bremen, Germany

Jorge Rabassa, Labaratorio de Geomorfología y Cuaternar, CADIC-CONICET, Ushuaia, Argentina

Vikram Unnithan, Department of Earth and Space Sciences, Jacobs University Bremen, Bremen, Germany

W0037703

SpringerBriefs in Earth System Sciences present concise summaries of cutting-edge research and practical applications. The series focuses on interdisciplinary research linking the lithosphere, atmosphere, biosphere, cryosphere, and hydrosphere building the system earth. It publishes peer-reviewed monographs under the editorial supervision of an international advisory board with the aim to publish 8 to 12 weeks after acceptance. Featuring compact volumes of 50 to 125 pages (approx. 20,000—70,000 words), the series covers a range of content from professional to academic such as:

- A timely reports of state-of-the art analytical techniques
- bridges between new research results
- snapshots of hot and/or emerging topics
- literature reviews
- in-depth case studies

Briefs are published as part of Springer's eBook collection, with millions of users worldwide. In addition, Briefs are available for individual print and electronic purchase. Briefs are characterized by fast, global electronic dissemination, standard publishing contracts, easy-to-use manuscript preparation and formatting guidelines, and expedited production schedules.

Both solicited and unsolicited manuscripts are considered for publication in this series.

Salomon Kroonenberg

The Changing Framework
of the Guiana Shield

 Springer

Salomon Kroonenberg ⓘ
Anton de Kom University of Suriname
Paramaribo, Suriname

ISSN 2191-589X   ISSN 2191-5903 (electronic)
SpringerBriefs in Earth System Sciences
ISBN 978-3-031-86333-2   ISBN 978-3-031-86334-9 (eBook)
https://doi.org/10.1007/978-3-031-86334-9

This Springer imprint is published by the registered company Springer Nature Switzerland AG
The registered company address is: Gewerbestrasse 11, 6330 Cham, Switzerland

If disposing of this product, please recycle the paper.

The changing framework of the Guiana Shield

The Voltzberg in Suriname, an Orocaima (2.0 Ga) granite pluton intruded in metagreywackes of the Maronian greenstone belt. Named after Friedrich Voltz (1828–1855) discoverer of the greenstone belt in the Guiana Shield. Photo by Ronny Bhoelai.

There is an alarming increase in things I
know nothing about
Don Tarling, Puerto Ayacucho, 1981

Dedicated to the memory of Theo Wong

Preface

The Guiana Shield, the northern half of the Amazonian Craton, is one of the most promising areas on earth for finding new gold deposits, and at the same time the hinterland of spectacular newly discovered offshore hydrocarbon fields. Yet, it is one of the least explored old nuclei of the earth's crust due to its tropical rainforest cover, deep weathering, poor accessibility, low population density, and multilingual colonial heritage. At the same time, the pristine rainforest that covers 70% of its surface area is one of the most untouched biodiversity hotspots in the world.

This book offers a review of the evolution of thoughts about the geological structure of the Guiana Shield from more than a century ago to the present. It is my intention to give credit to all those forgotten scientists and their splendid ideas by bringing their data to the limelight. I will show that many problems discussed at present have deep roots in the past: is the shield essentially an Archean craton reworked and remobilised by later Proterozoic events, or is it the result of continental accretion around pre-existing Archean nuclei? Is the tectonic development comparable to that in the Archean or to present-day plate tectonics? On the basis of many published maps and stratigraphic schemes, advances in stratigraphy, geochronology, geophysics, and structural geology I will show that many problems still have to be solved. The book is based on over a century of research in the basement of the Guiana Shield, and on my personal experiences from 1972 to the present, mainly in Suriname and Colombia. Most of the material has been published in local journals and conference proceedings. Almost all is grey literature, mostly not digitally available, published in five different languages, and illustrated with black-and white maps of unwieldy formats, poorly printed on transparency papers glued into clumsy >1000 page volumes. Relatively little has been published in the international journals.

This book is the elaboration of a talk given on August 21, 2023, at a workshop of the South American Exploration Initiative (SAXI) at Antraigues-sur-Volane, France. The SAXI project (saxiproject.org) is an initiative of AMIRA, an international non-profit organization based in Australia, who brought together a consortium of many mining companies to sponsor fundamental research in the Guiana Shield (2019–2024). I am grateful to the SAXI project staff, and especially to Mark Jessell as an inspiring project leader, for the close cooperation they forged between scientists from all Guiana Shield countries Brazil, Colombia, French Guiana, Guyana, Suriname and Venezuela, and interested international contributors, as well as for reviving the series of InterGuiana Geological Conferences after a standstill of over 40 years. This book, however, is not a SAXI publication, but an independent product for which I alone am responsible. I am grateful for the comments of two anonymous but very knowledgeable reviewers on an early version of the manuscript. There are no competing interests.

Paramaribo, Suriname   Salomon Kroonenberg

Contents

About the Author

Salomon Kroonenberg (1947) is an endowed professor of geology at the Anton de Kom University of Suriname and an emeritus professor of geology at Delft University of Technology, the Netherlands. He is also a guest researcher at Utrecht University. He obtained his Ph.D. in 1976 at the University of Amsterdam on high-grade metamorphism in southwestern Suriname. He was employed as a geologist at the Geological and Mining Service of Suriname from 1972 to 1978, as a lecturer of physical geography at the University College of Swaziland (1978–1979), and as a lecturer of remote sensing at the Centro Interamericano de Fotointerpretación CIAF in Bogotá, Colombia (1979–1982). In 1982 he was appointed as professor of geology at the Agricultural University Wageningen until his appointment in Delft in 1996. Since 2008 he has also had a honorary professorship at Moscow State University. His main interest at present is the Precambrian geology of the Guiana Shield, but he tackled many other subjects as well in his professional career, including Caspian Sea level change. He wrote over 150 scientific publications and 10 science books for non-professionals. He is past chairman of the Royal Netherlands Geological and Mining Society and of the Earth and Climate Council of the Royal Netherlands Academy of Arts and Sciences.

Chapter 1
Introduction

Abstract The Guiana Shield, the northern part of the Amazonian Craton, is roughly contained between the Atlantic Ocean, the Amazon River and the Orinoco River. This chapter discusses the rationale for a book on the main building blocks of the Guiana Shield, refers to the early notions on shields and cratons in general and to the nomenclature of the Guiana Shield in particular. A very short recent description of the geology of the Guiana Shield is given.

The Guiana Shield, the northern part of the Amazonian Craton, is roughly contained between the Atlantic Ocean, the Amazon River and the Orinoco River. Recently, it became the focus of intense geological and economical interest, for two reasons. In the first place, because according to the ex-CEO of Réunion Gold company David Fennell (pers.comm), Suriname and Guyana in the northern Guiana Shield are the best countries in the world to find new gold deposits.

Although the geological structure of the Guiana Shield in a pre-drift reconstruction forms a continuous belt with West-Africa (Fig. 1.1), it produces only one third of the gold of its counterpart. This is because about 70% of its surface area is covered with tropical rainforest (Hammond 2005), and deep tropical weathering and limited infrastructure hamper geological research. Moreover, the Guiana Shield straddles six countries: the three Guianas, Brazil, Venezuela and Colombia, and each country has studied the shield from its own perspective, with its own explorers, its own geologists, each publishing in its own language Spanish, Portuguese, English, French or Dutch. Its population density is only one tenth of that of West-Africa. That is why many companies consider the Guiana Shield as heavily underexplored and hence full of promises.

The second reason for the increased interest is that the northern Atlantic offshore of the Guiana Shield, the Guyana-Suriname Basin, has risen 'from obscurity to superpotential' (Cool and Viloria 2021). Oil companies have deeply invested in source-to sink studies to relate the properties of the reservoir systems to the drainage basins, denudation history and lithological characteristics of the Guiana Shield hinterland (Delhaye-Prat et al. 2024 and Fig. 1.2).

S. Kroonenberg, *The Changing Framework of the Guiana Shield*, SpringerBriefs in Earth System Sciences, https://doi.org/10.1007/978-3-031-86334-9_1

Fig. 1.1 In a pre-drift reconstruction the Maronian greenstone belts of the Guiana Shield continue in the Birimian greenstone belts in West Africa (Kioe-A-Sen et al. 2016)

Because of its inaccessibility, multicolonial and multilingual history it has taken a long time before something of a common concept on the geological structure of the Guiana Shield has emerged. While the plate tectonics revolution clarified many aspects of the origin of Phanerozoic orogenic belts, the question whether plate tectonics also acted in the build-up of the deeply eroded Precambrian nuclei of the continents is still under discussion. The origin of Archean and Paleoprotero-zoic greenstone belts and their relation to adjacent granulite-gneiss-belts has been intensely investigated especially in southern Africa and in the Canadian Shield, but has also spilled over to many other old cratons and shields. Here we present how these concepts influenced thoughts on the evolution of the Guiana Shield.

The purpose of this paper is to trace the development of ideas in the last one hundred years about the general structure and tectonic history of the Guiana Shield as a whole. I will discuss the larger Precambrian building blocks of the Guiana Shield, their age, origin and mutual relations in the broader framework of the shield. There-fore I have refrained from delving deep into local stratigraphies, and also paid little attention to smaller and late events such as dolerite dykes and sandstone covers which have less relevance for the broader picture. Mineral deposits will not be discussed in this text. I will draw more on review papers and published geological maps of the shield than on individual traveller's accounts, regional surveys and local reports. However, it is impossible to give credit to all authors that have written about the

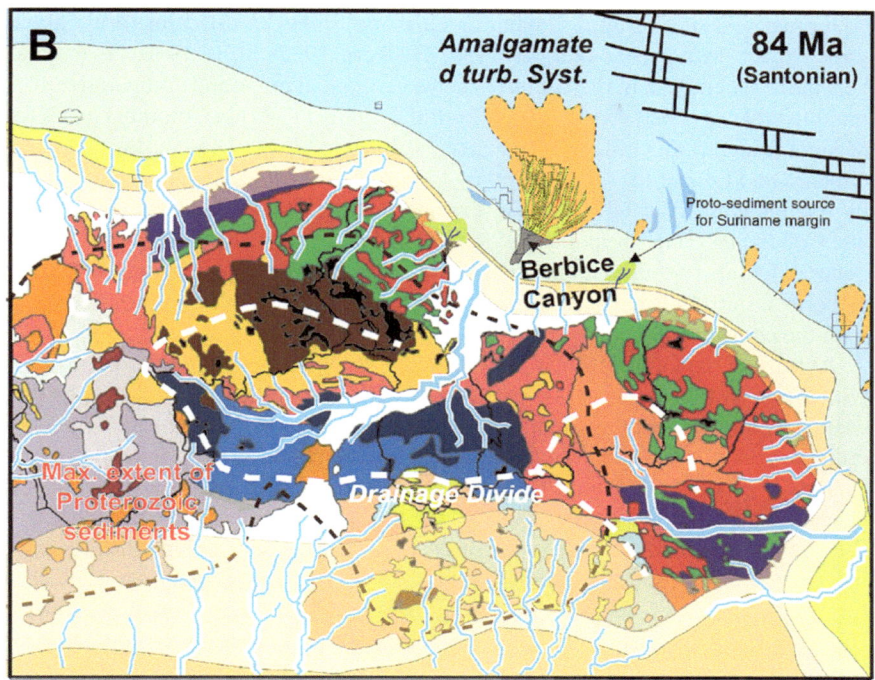

Fig. 1.2 Drainage basins and surface geology of the Guiana Shield for source and sink studies in the Suriname-Guyana basin in the Santonian (Delhaye-Prat et al. 2024). Reproduced with permission from Elsevier

geology of the Guiana Shield, this is just a modest review of hopefully the most important papers.

Whereas the name Guiana (Sp. Guayana) goes back at least to Sir Walter Ralegh's book *The Discoverie of the large, rich and Bewtiful Empyre of Guiana* (Ralegh 1596), the recognition of the Guiana highlands as a shield is not even a century old. Classifying the old nuclei of the continents as shields goes back to Eduard Suess, who in his *Das Antlitz der Erde* (*The Face of the earth*) in 1892 called the exposed Archaean surface of northeastern America the *Canadian Shield* (Suess 1906, p. 30). Likewise he recognized the Baltic Shield, and he called South America a shield surrounded by mountain ranges. The term *craton* is a much later development. Kober (1921) subdivided the present-day structure of the earth's crust in *orogens* and *kratogens*, while the latter term was shortened by Stille in 1939 to *craton* (Şengör 1999). A craton may include *platforms*, basement covered with essentially horizontal sedimentary rocks. In modern terms, the Guiana Shield forms the northern part of the Amazonian Craton, separated from the southern Guaporé or Central Brazilian Shield by the sedimentary Amazon basin.

In older literature the Guiana interior is referred to as massif guyanais (Argand 1924); Guayana Highlands (Dalton 1912; Liddle 1928) or Guiana Highlands

(IJzerman 1931). The term Guiana or Guayana Shield in different forms already turns up in older papers (Kugler 1936; Hedberg 1937; Stille 1940). It gradually gets more acceptance in the comprehensive geological maps and comparative stratigraphic tables as Guayana Shield by Rudolf Trümpy (1943) in Colombia and Victor Manuel López et al. (1942) in Venezuela, as Escudo Orenocoano or das Guianas by Oliveira and Leonardos (1940, cited by Issler, 1975) in Brazil, as Bouclier guyanais by Boris Choubert (1949, 1952) in French Guiana, as Guyanese schild by Schols and Cohen (1953) in Suriname and as Guiana Shield by Augusto Gansser (1954) in general. Others preferred calling it a Craton such as Edson Suszczynski (1970) and Roberto Issler (1975). We prefer to spell Guiana Shield with i to avoid confusion with the country Guyana.

At present the Guiana Shield is considered as a largely Paleoproterozoic massif, with Archean blocks on its western and eastern extremities: the Imataca block in eastern Venezuela and the Amapá block in northeastern Brazil (Fig. 1.3) The northernmost part of the Guiana Shield consists of the 1500 km long Paleoproterozoic Maronian granite-greenstone belt stretching from eastern Venezuela through the three Guianas into the state of Amapá in Brazil (Kroonenberg et al. 2016, 2019). We retain the name Maronian to mirror the Birimian in West-Africa. The greenstone belt and several high-grade metamorphic belts further south (Fraga et al. 2024) all formed between 2.26 and 2.08 Ga during the Trans-Amazonian Orogeny, coined by Hurley et al. (1967). The extensive 1.99–1.96 Ga felsic volcanic-granitoid Orocaima belt in the central part of the shield stretches from southern Venezuela through the Brazilian Roraima State and Guyana to southern Suriname (Reis et al. 2000, 2021; Fraga et al. 2024). It is bordered in the south and west by younger, Paleo- to Mesoproterozoic granitoid and high-grade metamorphic terrains in the southern Guianas, northern Brazil, southern Venezuela and eastern Colombia (Kroonenberg 2019).

The crystalline basement of the shield is overlain by a series of 1830 Ma horizontal sandstones and conglomerates, the Roraima Series, forming impressive platforms and escarpments up to 3000 m above sea level (Santos et al. 2003; Beyer et al. 2015), as well as some folded younger sandstone covers such as the Tunuí-Taraira sandstones overlying 1591 Ma granite (Ibáñez-Mejía 2010).

At least five suits of dolerite dykes intruded the Guiana Shield between 1980 and 200 Ma, including the 1783 Ma Avanavero LIP (Reis et al. 2013), the 1528 Ma Käyser alkalibasaltic dolerite swarm (De Roever et al. 2003a, b; Baratoux et al. 2019) and the 200 Ma CAMP dolerite (Nomade et al. 2000). The ~1980 Ma dolerites in northern Venezuela and northern Suriname and a ~400 Ma Devonian dolerite in Suriname still await publication. The NNW stretching 800 Ma (K–Ar) Tampok dolerite swarm of Delor et al. (2003) in southwestern French Guiana is situated exactly in line with the 200 Ma Apatoe dolerite further north in eastern Suriname, and therefore there are doubts about the correctness of its age.

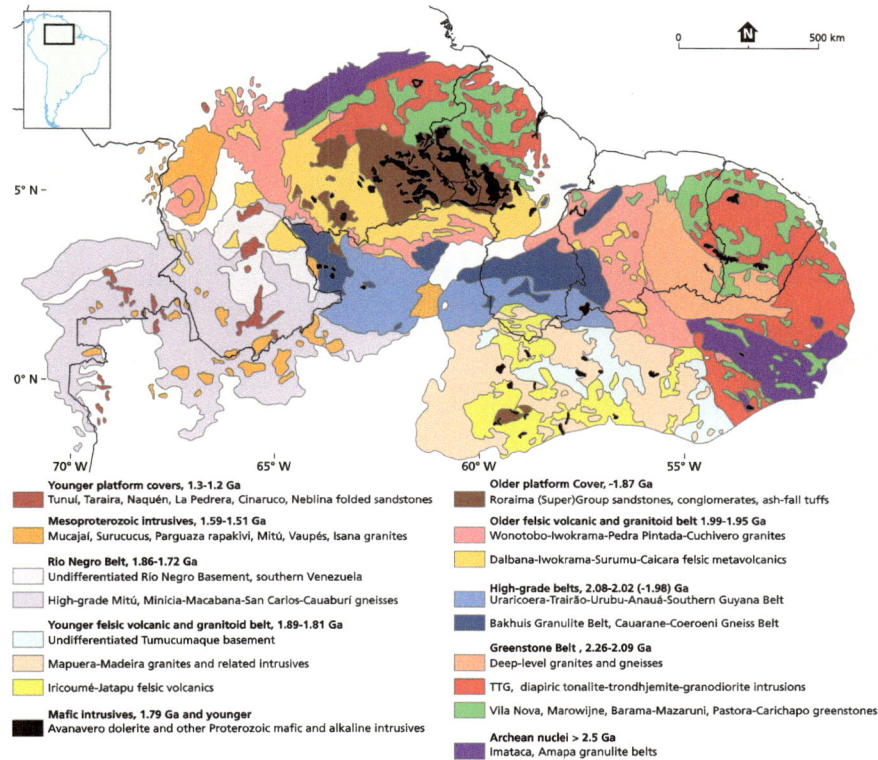

Younger platform covers, 1.3-1.2 Ga
Tunuí, Taraira, Naquén, La Pedrera, Cinaruco, Neblina folded sandstones

Mesoproterozoic intrusives, 1.59-1.51 Ga
Mucajaí, Surucucus, Parguaza rapakivi, Mitú, Vaupés, Isana granites

Río Negro Belt, 1.86-1.72 Ga
Undifferentiated Río Negro Basement, southern Venezuela

High-grade Mitú, Minicia-Macabana-San Carlos-Cauaburí gneisses

Younger felsic volcanic and granitoid belt, 1.89-1.81 Ga
Undifferentiated Tumucumaque basement

Mapuera-Madeira granites and related intrusives

Iricoumé-Jatapu felsic volcanics

Mafic intrusives, 1.79 Ga and younger
Avanavero dolerite and other Proterozoic mafic and alkaline intrusives

Older platform Cover, ~1.87 Ga
Roraima (Super)Group sandstones, conglomerates, ash-fall tuffs

Older felsic volcanic and granitoid belt 1.99-1.95 Ga
Wonotobo-Iwokrama-Pedra Pintada-Cuchivero granites

Dalbana-Iwokrama-Surumu-Caicara felsic metavolcanics

High-grade belts, 2.08-2.02 (-1.98) Ga
Uraricoera-Trairão-Urubu-Anauá-Southern Guyana Belt

Bakhuis Granulite Belt, Cauarane-Coeroeni Gneiss Belt

Greenstone Belt , 2.26-2.09 Ga
Deep-level granites and gneisses

TTG, diapiric tonalite-trondhjemite-granodiorite intrusions

Vila Nova, Marowijne, Barama-Mazaruni, Pastora-Carichapo greenstones

Archean nuclei > 2.5 Ga
Imataca, Amapa granulite belts

Fig. 1.3 A recent geological map of the Guiana Shield showing most units discussed in the text (Kroonenberg et al. 2016)

References

Argand E (1924) La tectonique de l'Asie. Extrait du Compte-rendu du XIII Congrès Géologique Internationale, Liège, pp 171–372

Baratoux L, Söderlund U, Ernst RE, de Roever E, Jessell MW, Kamo S, Naba S, Perrouty S, Metelka V, Yatte D, Grenholm M, Diallo DP, Ndiaye PM, Dioh E, Cournède C, Benoit M, Baratoux D, Youbi N, Rousse S, Bendaoud A (2019) New U–Pb Baddeleyite ages of mafic dyke swarms of the West African and Amazonian Cratons: implication for their configuration in supercontinents through time. In: Srivastava RK et al (eds) Dyke swarms of the world: a modern perspective. Springer Geology, pp 262–314. https://doi.org/10.1007/978-981-13-1666-1_7

Beyer SR, Hiatt EE, Kyser K, Drever GL, Marlatt J (2015) Stratigraphy, diagenesis and geological evolution of the Paleoproterozoic Roraima Basin, Guyana: links to tectonic events on the Amazon Craton and assessment for uranium mineralization potential. Precambr Res 267:227–249

Choubert B (1949) Géologie et pétrographie de la Guyane française. Office De La Recherche Scientifique Outre-Mer, p 143

Choubert B (1952) Carte géologique des trois Guyanes. Compte-rendu du 19ième congrès Géologique Internationale, Alger, section XIII, Fasc. XIV

Cool T, Viloria L (2021) Guyana-suriname Basin: from obscurity to super potential. World Oil

Dalton LV (1912) Venezuela. London, p 320

Delhaye-Prat V, Bourget J, Gaillot G, Gaillot J, Sapin F, Fillon C, J. Ye J, Wright T, Chaboureau AC , Buratti N, Magnier B, Belopolsky A, Bez M, Heumann MJ, Sullivan M, Mathieu J-P, Cole S, Ladner B, Bull J, Dal JA (2024) Tectono-sedimentary evolution of the Suriname margin in the cretaceous: a sequence-stratigraphic framework. Earth Sci Rev 253:104770

Delor C, Lahondère D, Egal E, Lafon JM, Cocherie A, Guerrot C de Avelar V (2003) Transamazonian crustal growth and reworking as revealed by the 1:500,000-scale geological map of French Guiana. Géologie de la France, No 2, 3, 4: 5–57

De Roever EWF, Kroonenberg SB, Delor C, Phillips D (2003a) The Käyser dolerite, a Mesoproterozoic alkaline dyke suite from Suriname. Géologie de la France, No 2, 3, 4:161–174

De Roever EWF, Lafon JM, Delor C, Rossi P, Cocherie A,.Guerrot C, Potrel A (2003b) The Bakhuis ultra-high temperature granulite belt : i petrological and geochronological evidence for a counterclockwise P-T path at 2.07–2.05 Ga. Géologie de la France 2–3–4:175–205

Fraga LM, Cordani U, Dreher AM, Sato K, Reis NJ, Nadeau S, de Roever E, Kroonenberg S, Maurer VC (2024) Early Orosirian belts of the central Guiana Shield, northern Amazonian Craton: U-Pb geochronology and tectonic implications. Precambr Res 407:107362. https://doi.org/10.1016/j.precamres.2024.107362

Gansser A (1954) The Guiana Shield (S. America). Eclogae Geol Helv 47:77–112

Hammond DS (2005) Tropical forests of the Guiana Shield. CABI Publishing, p 528

Hedberg HD (1937) Stratigraphy of the Rio Querecual section of Northeastern Venezuela. Bull Geol Soc Am 48:1971–2024

Hurley PM, de Almeida FFM, Melcher GC, Cordani UG, Rand JR, Kawashita K, Vandoros P, Pinson WH, Fairbairn HW (1967) Test of continental drift by comparison of radiometric ages. Science 157:495–500

Ibáñez-Mejía M (2010) New U-Pb geochronological insights into the Proterozoic tectonic evolution of Northwestern South America: the Mesoneoproterozoic Putumayo Orogen of Amazonia and implications for Rodinia reconstructions. MSc thesis University of Arizona, p 56

IJzerman R (1931) Outline of the geology and petrology of Surinam (Dutch Guiana). Kemink and Zoon N.V. (Utrecht), p 519

Issler RS (1975) Geologia do Cráton Guianês e suas possibilidades metalogenéticas. Anais Décima Conferência Geológica Interguianas, Belém, Brazil, pp 47–72

Kioe-A-Sen NME, van Bergen M, Wong TE, Kroonenberg SB (2016) Gold deposits of Suriname: geological context, production and economic significance. Neth J Geosci Geologie en Mijnbouw 95:429–445

Kober L (1921) Der Bau der Erde Gebr Borntraeger, Berlin, p 324

Kroonenberg SB (2019) The proterozoic basement of the western Guiana Shield and the Northern andes. In: Cediel F, Shaw RP (eds) Geology and tectonics of Northwestern South America. Springer, Frontiers in Earth Sciences, pp 115–192

Kroonenberg SB, de Roever EWF, Fraga LM, Reis NJ, Faraco MT, Cordani UG, Lafon J-M, Wong TE (2016) Paleoproterozoic evolution of the Guiana Shield in Suriname—a revised model. Neth J Geosci Geologie En Mijnbouw 95:491–522

Kroonenberg S, Mason PRD, Kriegsman LM, de Roever EWF, Wong TE (2019) Geology and mineral deposits of the Guiana Shield. In: Proceedings 11th inter Guiana geological conference, Paramaribo. Mededeling Geologisch Mijnbouwkundige Dienst Suriname vol 29, pp 111–115

Kugler HG (1936) Summary digest of the geology of Trinidad. AAPG Bull 20:1439–1453

Liddle RA (1928) The geology of Venezuela and Trinidad. Forth Worth, Texas, p 552

López VM, Mencher E, Brieman Jr JH (1942) Geology of Southeastern Venezuela. Bull Geol Soc Am 53:849–872

Nomade S, Théveniaut, H, Chen, Y, Pouclet, A, Rigollet, C, (2000) Paleomagnetic study of French Guyana Early Jurassic dolerites: hypothesis of a multistage magmatic event. Earth and Planetary Sci Letters 184: 155–168

Oliveira AI, Leonardos OH (1940) Geologia do Brasil. Com um mapa geológico do Brasil e parte dos países vizinhos. Comissão Brasileira dos Centenário de Portugal, Rio de Janeiro

Ralegh W (1596) The discoverie of the large, rich and bewtiful Empyre of Guiana. Facsimile 1966, Bibliotheca Americana, Cleveland, Ohio

Reis NJ, de Faria MSG, Fraga LM, Haddad RC (2000) Orosirian calc-alkaline volcanism and the Orocaima event in the Northern Amazonian Craton, Eastern Roraima State, Brazil. Revista Brasileira De Geociências 30(3):38–383

Reis NJ, Teixeira W, Hamilton MA, Bispo-Santos F, Almeida ME, D'Agrella-Filho MS (2013) Avanavero mafic magmatism, a late Paleoproterozoic LIP in the Guiana Shield, Amazonian Craton: U-Pb ID-TIMS baddeleyite, geochemical and paleomagnetic evidence. Lithos 174:175–195. https://doi.org/10.1016/j.lithos.2012.10.014

Reis NJ, Teixeira W, D'Agrella-Filho MS, Bettencourt JS, Ernst RE, Goulart LEA (2021) Large igneous provinces of the Amazonian Craton and their metallogenic potential in Proterozoic times. In: Srivastava RK, Ernst RE, Buchan KL, de Kock M (eds) (2022) Large igneous provinces and their plumbing systems. Geological Society, London, Special Publications, vol 518, pp 493–529

Santos JOS, Potter PE, Reis NJ, Hartmann LA, Fletcher IR, McNaughton NJ (2003) Age, source and regional stratigraphy of the Roraima Supergroup and Roraima-like outliers in northern South America based on U-Pb geochronology. Geol Soc Am Bull 115:331–348

Schols H, Cohen A (1953) De ontwikkeling van de geologische kaart van Suriname. Geol Mijnbouw 15:142–151

Şengör AMC (1999) Continental interiors and cratons: any relation? Tectonophysics 305:1–42

Stille H (1940) Einführung in den Bau Amerikas. Gebr. Borntraeger, Berlin, p 717

Suess E (1906) The face of the earth (Translation of Das Antlitz der Erde), vol II, Oxford, p 556

Suszczynski EF (1970) La Géologie et la Tectonique de la Plateforme Amazonienne. Geol Rundsch 59:1232–1253

Trümpy D (1943) PreCretaceous of Colombia. Geol Soc Am Bull 54:1281–1304

Chapter 2
The Early Days: Geosynclines and Granitization

Abstract The Guiana Shield first appeared as a recognisable geological entity on 19th century and early 20 century geological maps of South America and Brazil. From the 1950ies onward concerted efforts by the three Guianas resulted in joint maps. Due to the difficult outcrop conditions, lack of fossils and of a reliable chronology, stratigraphies varied widely and orogenic events were largely described on the basis of concepts developed in Phanerozoic belts.

The Guiana Shield first appeared as a recognisable geological entity on the 1856 geological map of South America by the Austrian geologist Franz Foetterle (Fig. 2.1). Foetterle did not visit South America himself, but compiled a lot of data from different sources (Foetterle 1856, cf. Lobitzer and Kadletz 2005; Rossi 2014; Schobbenhaus 2014, Bartoletti 2022).

John Casper Branner (1919) presents a coloured geological map of Brazil, with Archean for the whole Guiana Shield (without using that name) and some green spots for the supposedly Cretaceous Roraima sandstone plateau (Fig. 2.2). He also cites details about earlier Brazilian maps.

In the interesting map of Gondwana by Émile Argand (1924, Fig. 2.3), a staunch supporter of Wegener, the Guiana Shield and the other South American shield areas are grouped with those from western Africa in the same 'pli de fond' III, a huge anticlinal fold brought about by compression between the continents. It is the first time a collision between South America and Africa was envisaged, a concept now largely adopted to explain the Trans-Amazonian Orogeny ~2 Ga ago.

The first geological map of Brazil by Brazilian geologists Oliveira and Leonardos (1940, Fig. 2.4) showed also the southern part of the Guianas, but again with only the Roraima formation and the basement (Lopes 2022).

In 1947 the first Director of the Geological and Mining Service of Suriname Hendrik Schols was asked by the Geological Society of America (GSA) to contribute to the Geological map of South America which was going to be published in 1951 (Fig. 2.5). Schols derived almost all data about the Suriname basement for the GSA map from the monumental work of IJzerman (1931), which appears clearly on the

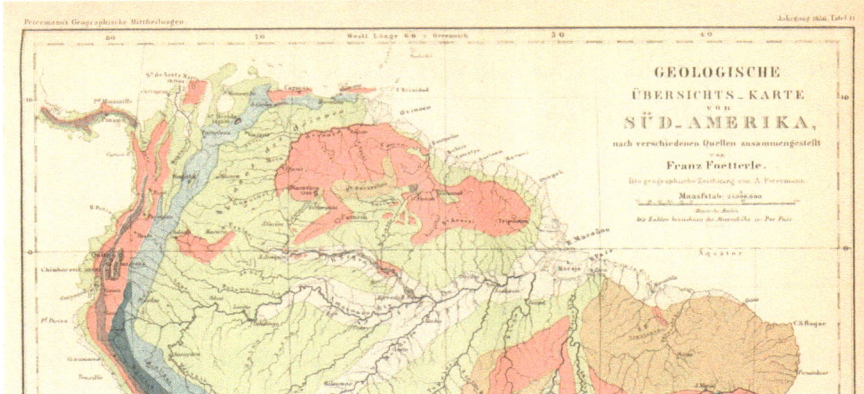

Fig. 2.1 The Guiana Shield on the first geological map of South America (Foetterle 1856, Petermanns Mitteilungen 2, Karte 11)

Fig. 2.2 Part of the coloured geological map of Brazil map by Branner (1919)

map: mainly granites with greenstones. The subdivision and legend symbols were prescribed by the Society and were submitted as in Fig. 2.6 (Schols and Cohen 1953).

However, the authors of the Geological Map of South America apparently were not aware that the first detailed geological map of the Guiana Shield had already been published by Choubert (1949). He was a Russian-born French geologist, who as an employee of the Office Scientifique d'Outre-mer laid the framework for the geology of French Guiana between 1946 and 1960. Before that, he had worked as a geologist in Gabon, and already then he was a great supporter of Wegener's theory of continental drift, against the mainstream consensus in those years. In 1935

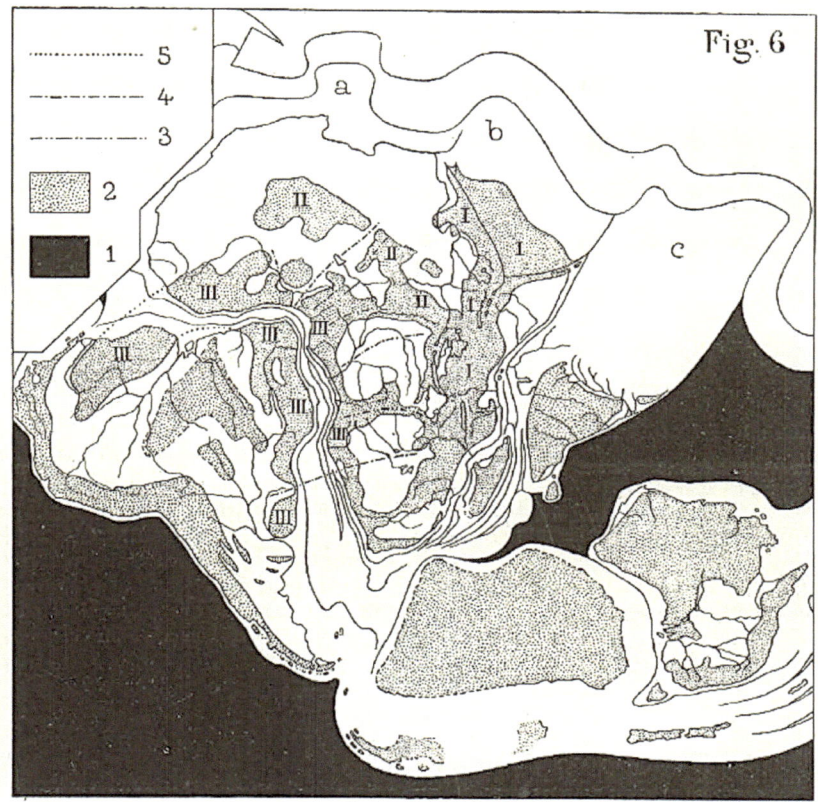

Fig. 2.3 The South American and African massifs are huge anticlinals formed by the collision of those continents (Argand 1924)

he published a convincing map of the continental fit of the continents around the Atlantic Ocean (Kornprobst et al. 2018).

The legend of Choubert's (1949) map (Figs. 2.7 and 2.8) in the shield area shows an Antecambrian crystalline metamorphic basement with greenstones (roches vertes), granitic intrusions at the base, followed by Imataca Quartzites, Paleozoic rocks and the Roraima series on top, then still thought to be Cretaceous in age. The greenstones in Venezuela, British Guiana, Suriname, French Guyana and the Amapá state in Brazil are indicated with a special signature, based on the works of Liddle (1928) in Venezuela, Harrison (1908) and Bracewell (1947) in British Guiana, IJzerman (1931) in Suriname, his own work in French Guiana, and Vélain (1881) for Amapá. Amapá was in Vélains time still claimed by France (Levat 1898). Choubert's subdivision foreshadows the classic threefold subdivision of the Archean greenstone belts in southern Africa. His interpretation of the events is summarized as that they underwent 'several granitizations and were largely metamorphosed'. These folded sequences

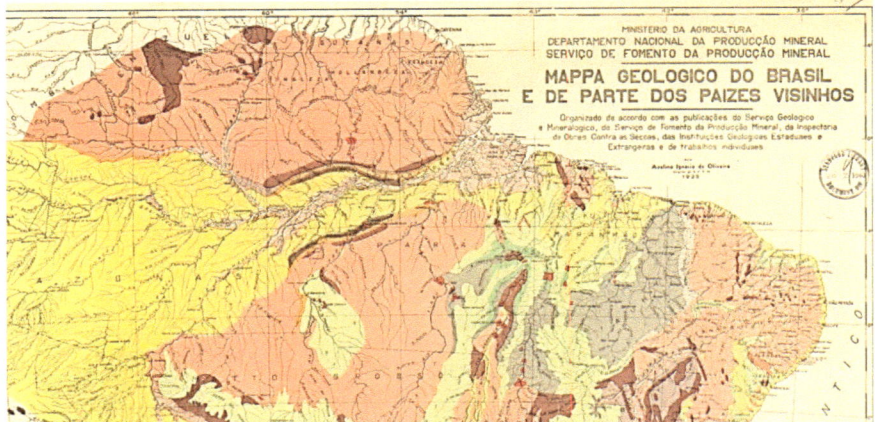

Fig. 2.4 Part of the 1940 geological map of Brazil by Oliveira and Leonardos (after Lopes 2022)

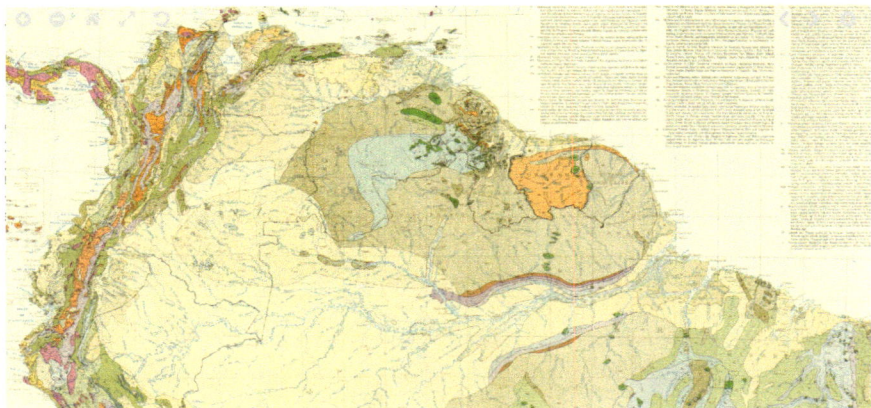

Fig. 2.5 Detail of geological map of South America, 1950, including Roraima and Schols' contributions

are overlain by Paleozoic rocks and by the Roraima series on top. The geological map of Suriname by Martin (1888), who first defined a greenstone belt in the Guiana Shield based on the work by Friedrich Voltz in Suriname in 1853–1855 (Kroonenberg 2020, 2022) was probably still unknown to him.

The stratigraphic table of Chouberts map starts with lower quartzites at the base in French Guiana which do not appear separately on the Guiana Shield map, only at Mont d'Argent on the coloured French Guiana map in the same volume. The sequence of greenstones followed by conglomeratic and quartzites in each of these units is uphold up to the present day. The main units are separated by four period of folding, granite intrusions and peneplanization.

VI.	*Latest Deposits*		
10.	Coastal, fluviomarine alluvium	Holocene	Qm
9.	Continental alluvium	⎰	
8.	Laterite Ironstone and Bauxite	Pleistocene to Miocene	Qc or T
V.	*Youngest Basic Intrusives*		
7.	Diabase-gabbro-intrusions	post Triassic	Kib?
IV.	*Roraima (Kaieteurian) formation*		
6.	Sandstones and conglomerates	Triassic	Trc
III.	*Younger intrusives*		
5.	Quartzporphyries	Triassic?	Tri?
II.	*Older Intrusive Rocks*		
4.	Gabbro-norite intrusions	⎰	
3.	Granitodiorite intrusions (batholiths)	pre-Cambrian	pCi
I.	*Basement Rocks (Metamorphic-Series)*		
2.	Paraschists chiefly, perhaps some granite-gneiss	early pre-Cambrian	epC
1.	Basic orthoschists: amphibolites and hornblendeschists	pre-Cambrian basic intrusive	pCb

Fig. 2.6 Legend for the Guiana Shield as required for the Geological Map of South America (Schols and Cohen 1953)

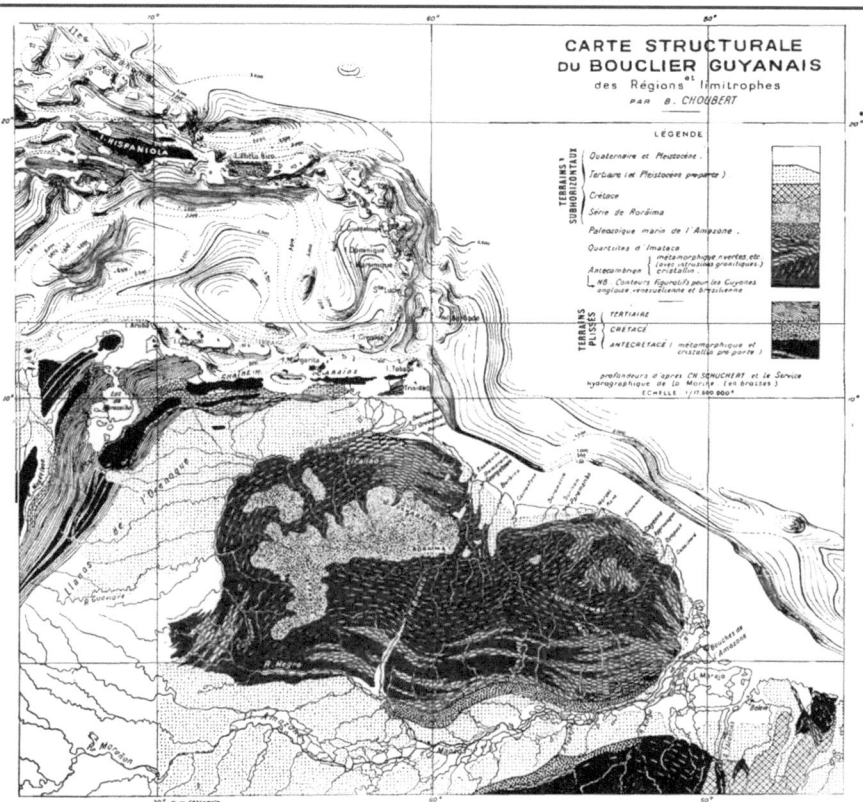

Fig. 2.7 'Structural' map of the Guiana Shield by Choubert (1949). Compare with Fig. 3.4

GUYANE VÉNÉZUÉLIENNE.	GUYANE ANGLAISE.	GUYANE HOLLANDAISE.	GUYANE FRANÇAISE.
Dépôts du delta de l'Orénoque.	Demerara-clay.	Vase bleue.	Vase bleue.
Dépôts continent. des Llanos.	White Sand Series sup. (Berbice form.).	Sables blancs.	Série des sables sup.
	Période d'érosion avec formation de bauxite.		
	Berbice form. inf.	Sables blancs inf.	Série des sables inf.
	Longue période d'érosion, pénéplaine avec formation de bauxite.		
	Intense activité volcanique, plusieurs venues de dolérites.		
Série de Roraïma.	Série de Roraïma.	Série de Roraïma.	?
	Plissements, longue période d'érosion, pénéplaine ante-Roraima.		
Série d'Imataca.	?	?	?
	Activité volcanique. — Rhyolites.		
	Plissements, venues granitiques, longue période continentale.		
?	Schistes.	Schistes à séricite.	Schistes de l'Orapu conglomérats et quartzites.
	Plissements, venues granitiques, longue période continentale.		
	Intense activité volcanique, diorites, gabbros et laves.		
Roches vertes (Pastora series).	Volcanic series.	Roches vertes et schistes inf.	Roches vertes et schistes inf.
	Plissements, venues granitiques.		
?	?	?	Quartzites inf.

Fig. 2.8 Stratigraphy of the northern Guiana Shield (Choubert 1949 p. 106 eV)

Hendrik Schols also organised the First Inter-Guiana Geological Conference (IGGC) in Paramaribo from September 23 to October 3, 1950. Boris Choubert attended as a representative of French Guiana, and Smith Bracewell as Director of the British Guiana Geological Survey. The event was followed in 1951 by the second IGGC organised by Choubert in Cayenne. In the meetings they agreed to construct a 1:1,000.000 geological map of the three Guianas, on the basis of joint discussions about the stratigraphy and joint field trips, and taking in account all previous maps since 1888. Choubert presented the joint map in 1952 at the 19th International Geological Conference in Alger (Fig. 2.9), but it received little attention. The third IGGC took place in Georgetown in 1953 (Schols and Cohen 1953; Kroonenberg 2019).

The legend of the joint map of the three Geological Surveys already showed much more detail than Chouberts 1949 map, including four units in the 'Antécambrien' the part which we now would call the greenstone belt. It starts with paragneisses and granitoid gneisses, in which we now recognise among others the Bartica gneiss in Guyana, the Kanuku and Bakhuis granulites in respectively Guyana and Suriname, and several terrains in central French Guiana. They are followed by mafic volcanics and two series of metasedimentary formations, Ga-Kaba/Bonidoro and

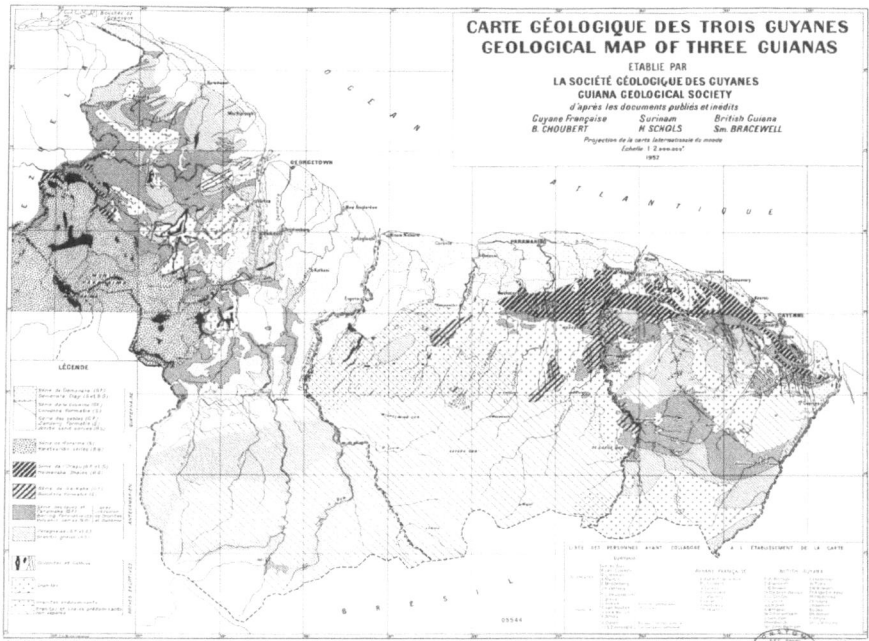

Fig. 2.9 The geological map of the three Guianas (Choubert, 1952) presented in Alger in 1952 (*source* BRGM)

Orapu/Haimaraka whose denominations have considerably changed during history, but still make sense in the light of the threefold successions commonly found in greenstone belts all over the world. The higher metamorphic grade of the gneisses is apparently taken as evidence for their older age, partly also based on Choubert's own 1949 work on the Île de Cayenne in French Guiana, even now considered to be one of the the oldest rocks within the Paleoproterozoic greenstone belts (Vanderhaeghe et al. 1998). So far for the greenstone/gneiss controversy in those years. The age of the Roraima sandstones is still unresolved. Igneous rocks are mapped in a separate category.

Augusto Gansser was a prodigious Swiss geologist who wrote a number of classical papers about the geology of South America and the Himalayas. He lived in Colombia from 1938–1946 and 1947–1949 on Trinidad, employed by Shell. He travelled extensively in the Colombian part of the Shield and held an expedition in British Guiana to Mount Roraima with support of Smith Bracewell, which led to his seminal paper *The Guiana Shield* (1954), the first geological paper dedicated exclusively to the Guiana Shield (Fig. 2.10). His map shows only the Roraima Formation and the Guiana Shield Complex. In a note added in proof in the paper he regretted he had not been able to consult Chouberts 1949 book, otherwise he probably would have given much more detail in his map.

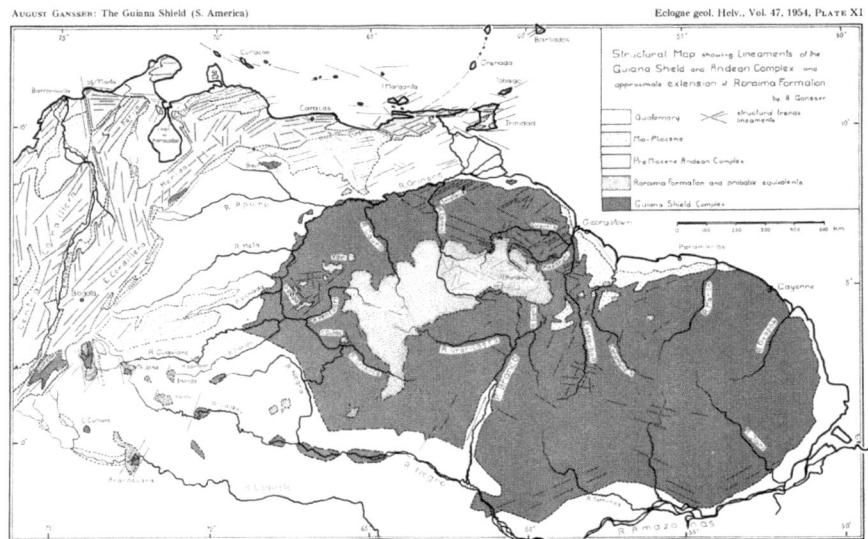

Fig. 2.10 The Guiana Shield according to Gansser (1954)

In his paper he presents a simple fourfold subdivision of the shield: he distinguishes Subrecent to recent deposits, the Roraima Formation, the Volcanic Group and the Basement Group. From the basement he described mainly granitoid and syenitic rocks from Colombia and Venezuela, as well as the Imataca itabirites, while the Volcanic Group refers to what we now call the greenstone belts of northern Guyana. In it he distinguishes a lower part, represented by volcanic rocks and its related tuffs, and an upper part grading from tuffs to shaly sediments (Haimarakka shales).

The main value of his paper, however, is an elaborate description of the stratigraphy of the Roraima Formation. From his own observations in British Guiana and adjacent parts of Venezuela and Brazil he distinguished (a) a basal member, including the basal conglomerates if present, (b) a middle member, characterized by frequent intercalations of jasper beds, and (c) an upper member, comprising the thickbedded sandstone horizons forming the spectacular cliffs of the highest table mountains. He was puzzled by the shear volume of the sediments and the absence of fossils or any other indicator of its age. He preferred a Cretaceous age on the basis of their resemblance to Cretaceous sandstone formations in the bordering Andean range. Even after convincing geochronological data establishing its age around 1700 Ma became available (Snelling 1963; Priem et al. 1968, 1973), he did not abandon the idea that part of the sequence might be younger, even Cretaceous indeed (Gansser 1974). Augusto Gansser died in 2012 at the age of 101 and was cremated on his explicit wish with his hammer beside him instead of flowers.

The 4th Inter-Guiana Geological Conference was held in Cayenne in 1957, and for the first time also Brazil was represented, by Luciano Jacques de Moraes with a paper on mineral resources of Amapá (Moraes 1959). Choubert presented a new geological map of French Guiana, partly also based on the report on the southern part of the country by Aubert de la Rue (1953), which is almost completely underlain by greenstones and granites. At the base he placed the amphibolites of the Île-de-Cayenne: even at present they still are considered among the oldest rocks in the Guiana Shield. The mafic to felsic volcanics on top of that, including associated intrusives, he called Paramaka, also still in use both in French Guiana and Suriname. They are followed by the Bonnidoro series, corresponding the Greywacke Formation of IJzerman (1931) in Suriname, and on top of that the clayey schists and quartzites of the Orapu series, already defined on his 1949 map. However, in Suriname this nomenclature has led to considerable confusion. Clear contacts between the formations have never been found in the field, nor in diamond drill cores, and one senses the lack of sedimentological information and radiometric data, but even today not all stratigraphic problems have been solved. Guyana had made its own stratigraphy which was still difficult to reconcile with Suriname and French Guiana.

At the 5th IGGC in 1959 in Georgetown the Deputy Director of the Geological and Mining Service of Suriname Guus Doeve (1961) discussed these problems in rock correlation in Suriname, especially the correlation between Paramaka volcanics and the sedimentary formations Bonnidoro, Rosebel, Orapu and their relation to granite intrusions. Also the hosts from Guyana presented their newest national map and a stratigraphic table (McConnell and Dixon 1961). Choubert stressed the importance magmatism in the chronology of the Shield.

For the first time Venezuela came with a large delegation to the IGGC conference, and its top geologists Cecilia Martín-Bellizzia and her husband Alirio Bellizzia presented a paper on the stratigraphic column of Venezuelan Guayana, in which many similarities with the three Guianas became apparent (Martín-Bellizzia and Bellizzia 1961). At the end of the conference the progress in drafting an International Geological map of South America and a map of the Guianas were discussed, as well as the forthcoming *Lexique Stratigraphique Internationale* for the Guianas. Both projects were hampered by the difficulty of setting up a reliable stratigraphy and correlating rock units across the countries. In the latter booklet the discrepancies between the stratigraphic nomenclature in the three countries became evident (Choubert et al. 1962).

In 1963, at the 6th InterGuiana Geological Conference in Belém, Brazil, Leon O'Herne, geologist at the Geological and Mining Service of Suriname, presented a new black and white map of the three Guianas, that showed the progress in knowledge after the 1952 map presented in Alger (Figs. 2.11 and 2.12). O'Herne (1969a) especially emphasized the importance of aerial photographs in distinguishing different rock units, an approach already advocated by Choubert in French Guiana in 1957 (Choubert 1957). In 1966, at the 7th IGGC in Paramaribo he presented a detailed 1:1,000.000 photogeological map of Suriname in colour (O'Herne 1969b).

In the geological map of South America drawn by the eminent Russian geologist Viktor Yefimovich Khain in 1967 (Khain 1971) the Guiana Shield stands out with remarkable detail (Fig. 2.13), showing not only the Imataca belt, the greenstone belts and the Roraima series, but also structures in the Brazilian Amazones and

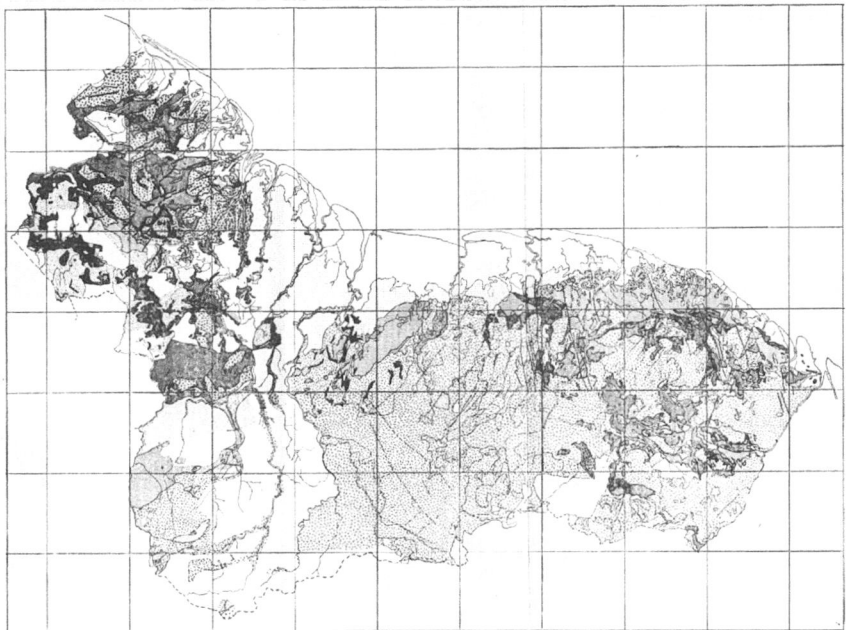

Fig. 2.11 Geological map of the three Guianas (O'Herne 1966)

the deviant structure of the western part of the shield in Colombia and Venezuela, which was later to be recognized as the Rio Negro block. There are similarities with the map of his erstwhile countryman Choubert (1949), and he cites Choubert's first radiometric datings from the Île de Cayenne, as well as those of Mc Connell in Guyana and Priem in Suriname (see next chapter). In his book he describes the main units in detail, considering the greenstone belts as ultrageosynclinal features. During a meeting in Montevideo in 1967 for the tectonic map of South America Khain had met the foremost Brazilian geologist Fernando de Almeida and his staff, who also may have been the source of the Brazilian information to which Venezuelan and Guianan geologists up to then apparently had no access.

Up to the 1960ies, little thought was given to the structural and tectonic evolution of the Guiana Shield. Most efforts were dedicated to establish correlations between the different geological units in the Guianian countries. The stratigraphic schemes were modeled according to those for the Phanerozoic, and interpreted as volcanic and sedimentary successions in geosynclinal basins, interrupted and deformed by orogenic cycles evidenced by unconformities and intrusive magmatism. But the lack of of a fossil-based chronology and of clear contacts between the formations led to unsatisfactory and controversial insights in the tectonic and chronological development of the shield.

LEGEND

British Guyana	Surinam	French Guyana		
Demerara form.	Demerara serie	série de Demerara	marine and fluviomarine deposits of Young Coastal Plain: bluish grey clays and silts; sands; sands or sandy ridges containing shells	Holocene
Coropina form.	Coropina serie	série de Coswine	tidal flat, clay delta and lagoon deposits of Old Coastal Plain: mottled silty clays and loams with intercalated coarse sands; fine grained sands (old offshore bars)	Plistocene
Berbice (white Sands) form.	Coesewijne serie	série Détritique de Base	continental detritic deposits with pluvial wash-out and creep, continental-deltaic and littoral deltaic deposits: coarse sands, sandy clays, clays rich in kaolin, gravels, lignites	Plio-Plistocene
Bauxite	Bauxiet		lowland-bauxite deposits on Eocene surface outcropping or with overburden of young sediments	U. Eocene
Younger deposits overlapping Takutu form.?			clays, sands, laterite	U. Tertiary – Recent
Takutu form.			shales and sandstones with plant remains and ostracods	L. Tertiary – U. Cretaceous
Younger Basic Intrusives	Jonge Basische In-en Extrusiva	Doléritas	dolerite (gabbro) sills and dykes	Mesozoic?
Roraima form.	Roraima form.		sandstones, conglomerates, quartzites, (shales, rhyolitic tuffs)	L. Paleozoic – U. Precambrian
Younger Granites Bartica gneisses, and-granite			biotite and hornblende (biotite) granites and – gneisses, muscovite granites, muscovite-biotite gneiss	U.-M. Precambrian
part of Younger Granites? Kartabu	Graniet 3	Granites Caraïbes	biotite and hornblende granites and gneisses, muscovite granite, muscovite-biotite gneiss (B.G.)	..
part of Younger Granites and of Bartica gneiss?	Graniet 2	Granite Guyanais	mostly (hornblende) – biotitite quartzdiorites, and gneisses, locally with microcline	M. Precambrian
Barama-Group and part of Mazaruni Group	Paramaka form.	syst. de Paramaka	metasediments (shales, schists, phyllites, argillites, quartzites, conglomerates) metavolcanics (lavas, tuffs, agglomerates, breccias); metam. basic intrusives, amphibolites, epidiorites)	L. M. Precambrian
Amphibolitic rocks of Bartica and Barama, basic volcanics or in-trusives of Mazaruni Group	Oude Basische In-en Extrusiva (including amphibolitic rocks of Paramaka)	Diorite, Gabbro, Amphibolite (of Granites Hyléens and Paramaka)	metam. diorites, gabbros, norites, pyroxenites, dolerites, amphibolites; correlation uncertain	several intrusive periods L.-U. Precambrian?
Kuyuwini Group	rhyolieten	rhyolites	rhyolites, andesites and tuffs (B.G.) rhyolites (Sur. & Fr.G.): not indicated on map	several extrusive periods L. Precambrian? Paleozoic
South Savanna Granite		synorogenic with Granites Hyléens?	biotite – (muscovite) – granite (B.G.) diorites, quartzdiorites, granodiorites, amphibole-biotite gneiss (Fr. G.)	L. Precambrian
Marudi Group	part of highly metam. rocks of Adampada – Falawatra region?		biotite-schists and gneisses with metaquartzites	L. Precambrian
Kanuku Group	Highly metam. rocks of Adampada – Falawatra region	syst. Ile de Cayenne?	granulites, biotite gneisses, pyroxene-sillimanite- and garnet-gneisses (B.G. & Sur.) (mica-garnet-) amphibolites, gneisses, (pyrox.) quartzites, metam. gabbro and diorite (Fr. G.)	L. Precambrian
			Na and (K) metasomatism?	
			K mesomatism?	

Fig. 2.12 Legend map Guiana Shield (O'Herne 1966)

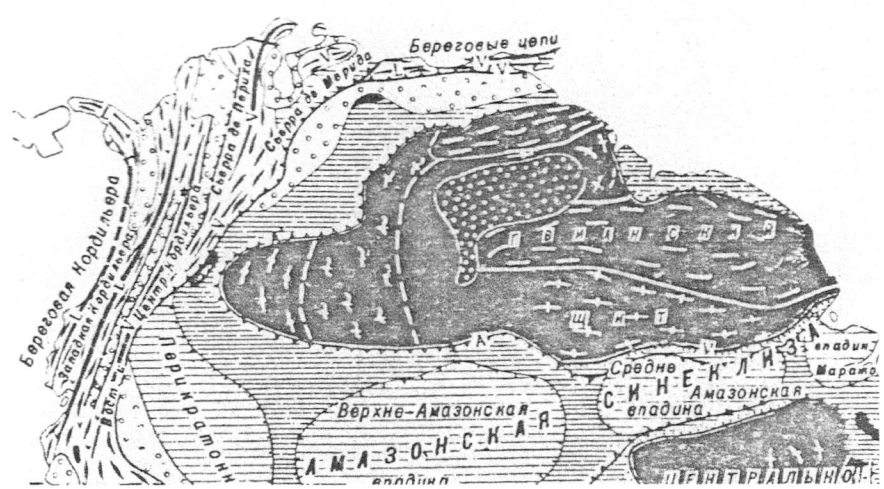

Fig. 2.13 Detail of the Tectonic scheme of South America map (Khain 1971; map dated 1967)

References

Argand E (1924) La tectonique de l'Asie. Extrait du Compte-rendu du XIII Congrès Géologique Internationale, Liège, 171–372

Aubert de la Rue E (1953) Reconnaissance géologique de la Guyane française méridionale 1948–1949–1950. Office de la Recherche Scientifique Outre-Mer, Paris, p 127

Bartoletti T (2022) Cartography in translation between Ouro Preto and Gotha, c.1850–1860. Imago Mundi 74:63–81. https://doi.org/10.1080/03085694.2022.2042126

Bracewell S (1947) The geology and mineral resources of British Guiana, Imp. Inst 45(1), London

Branner JC (1919) Outlines of the geology of Brazil to accompany the geologic map of Brazil. Bull Geol Soc Am 30:189–338, with coloured map

Choubert B (1949) Géologie et pétrographie de la Guyane française. Office De La Recherche Scientifique Outre-Mer, p 143

Choubert B (1952) Carte géologique des trois Guyanes. Compte-rendu du 19ième congrès Géologique Internationale, Alger, section XIII, Fasc. XIV

Choubert B (1957) Essai sur la morphologie de la Guyane. Mémoire pour servir à l'explication de la Carte Géologique de la France, Département de la Guyane Française, Paris, p 48

Choubert B, Lelong F, van Eijk HTL, Dixon CG, Bleackley D (1962) Lexique Stratigraphique International, vol V. Amérique Latine, Fasc. 10b Guyanes. Centre National de la Recherche Scientifique, p 77

Doeve G (1961) Problems in rock correlation in Surinam. In: Proceedings of the Fifth Inter-Guiana Geological Conference, Georgetown, British Guiana, pp 47–56

Foetterle F (1856) Geologische Übersichts-Karte von Süd-Amerika nach verschiedenen Quellen zusammengestellt. Map 11 in Petermanns Mitteilungen 2

Gansser A (1954) The Guiana shield (S. America). Eclogae Geol Helv 47:77–112

Gansser A (1974) The Roraima problem (South America). Verhandlungen Naturforscher Gesellschaft Basel 84:80–100

Harrison JB (1908) Geology of the goldfields of British Guiana. London, p 320

IJzerman R (1931) Outline of the geology and petrology of Surinam (Dutch Guiana). Kemink and Zoon N.V. (Utrecht), p 519

Khain VE (1971) Хайн В Е Региональная геотектоника—Северная и южная Америка, Антарктида и Африка. Издательство Недра, Москва, 1971, 548 стр. (236–348) (Regional geotectonics: North and South America, Antarctica and Africa.) Nedra editions, Moscow 548, pp 236–348

Kornprobst J, Àbalos B, Barbey P, Boullier AM, Burg JP, Capdevila R, Claesson S, Cordani U, Corrigan D, Gabrielsen RH, Gil-Ibarguchi JI, Johansson Å, Letsch D, Le Vigouroux P, Upton B (2018) Boris Choubert: unrecognized visionary geologist, pioneer of the global tectonics. BSGF Earth Sci Bull 189:7. https://doi.org/10.1051/bsgf/2018006

Kroonenberg S (2019) History of the inter Guiana geological conferences. In: Proceedings XI interguiana geological conference 2019: Paramaribo, Suriname. Mededeling Geologisch Mijnbouwkundige Dienst Suriname 29: iii–iv

Kroonenberg SB (2020) De man van de berg. Friedrich Voltz (1828–1855), jonggestorven natuuronderzoeker in Suriname. Walburg pers, Zutphen, p 320

Kroonenberg S (2022) Friedrich Voltz (1828–1855), discoverer of the Maronian greenstone belt in the Guiana shield. In: Proceedings 12th inter Guiana geological conference, Georgetown, Guyana, p 86

Levat MED (1898) Guide pratique pour la recherche et l'exploration de l'or en Guyane française. Paris, p 243

Liddle RA (1928). The geology of Venezuela and Trinidad. Forth Worth, Texas, p 552

Lobitzer H, Kadletz K (2005) Franz Foetterles Südamerikakarte. In: Schedl A, Hofmann T (eds) Grenzenlos. Berichte der Geologische Bundesanstalt Wien 62, pp 9–10

Lopes MM (2022) The geological map of Brazil 1938–1940. Earth Sci History 41:36–350

Martin K (1888) Aanteekeningen bij eene geognostische overzichtskaart van Suriname. Tijdschrift Koninklijk Nederlands Aardrijkskundig Genootschap 5:444–453

Martin-Bellizzia C, Bellizzia A (1961) Columna estratigráfica provisional de la Guayana Venezolana. In: Proceedings of the fifth inter-Guiana geological conference, Georgetown, British Guiana, pp 29–31

McConnell RB, Dixon CG (1961) Presentation of the revised geological map of British Guiana. In: Proceedings of the fifth inter-Guiana geological conference, Georgetown, British Guiana, pp 17–30

Moraes LJ (1959) Bauxites et autres richesses minières du Territoire Fédéral d'Amapa, Brésil. In: Communications présentées à la Quatrième Conférence Géologique des Guyanes. Cayenne, pp 93–97

O'Herne L (1966) Some remarks about a recent geological map of the three Guianas. In: Anais da VI Conferência geológica das Guianas 1963, Belém, pp 179–183

O'Herne L (1969a) A new interpretation of the stratigraphy of Surinam. Mededelingen Geologisch Mijnbouwkundige Dienst Suriname 20:9–12

O'Herne L (1969b) A photogeological study of the basal complex of Suriname. Mededelingen Geologisch Mijnbouwkundige Dienst Suriname 20:53–149

Oliveira AI, Leonardos OH (1940) Geologia do Brasil. Com um mapa geológico do Brasil e parte dos países vizinhos. Comissão Brasileira dos Centenário de Portugal, Rio de Janeiro

Priem HNA, Hebeda EH, Boelrijk VRH, Verdurmen EAT (1968) Isotopic age determination on Surinam rocks, 4: ages of basement rocks in North-Western Surinam and of the Roraima Tuff at Tafelberg. Geol Mijnbouw 47:191–196

Priem HNA, Boelrijk NAIM, Hebeda EH, Verdumen EAT, Verschure RH (1973) Age of the Precambrian Roraima formation in northeastern South America: evidence from isotopic dating of Roraima pyroclastic volcanic rocks in Suriname. Geol Soc Am Bull 84:1677–1684

Rossi P (2014) Celebración del centenario de la CCGM en Villa de Leyva, Colombia 21 de Julio 2014. In: Memoria geological map of South America workshop, pp 35–64

Schobbenhaus C (2014) CGMW 50 years of activities in South America. In: Memoria geological map of South America workshop, Villa de Leyva, Colombia, pp 65–74

Schols H, Cohen A (1953) De ontwikkeling van de geologische kaart van Suriname. Geol Mijnbouw 15:142–151

Snelling NJ (1963) Age of the Roraima formation, British Guiana. Nature 198:1079–1080

Vanderhaeghe O, Ledru P, Thiéblemont D, Egal E, Cocherie A, Tegyey M, Milési JP (1998) Contrasting mechanism of crustal growth: geodynamic evolution of the Paleoproterozoic granite–greenstone belts of French Guiana. Precambr Res 92:165–193

Vélain C (1881) Notes géologiques sur la Guyane française, d'après les collections recueillies par le docteur Crevaux. Bull Soc Géol France 3(IX):396

Chapter 3
The Dawn of Geochronology in the Guiana Shield

Abstract In the 1960ies, the first systematic K-Ar and Rb-Sr radiometric ages were obtained both in the Guianas, Venezuela and Brazil. They established the largely Archean and Paleoproterozoic ages of the Guiana Shield and confirmed the concept of continental drift by comparison with West-African chronology.

The first radiometric age from the Guiana Shield was already obtained in 1941: an euxenite, an uranium-bearing mineral from a mylonite zone in the Kanuku gneisses gave an age of 1090–1160 Ma (McConnell and Dixon 1961; Snelling and McConnell 1969). This is an amazing result, as it coincides with the ages of 1300–1100 Ma obtained for the K'Mudku or Nickerie event, a shield-wide event characterized by mylonitization along ENE-WSW shear zones and thermal resetting of K–Ar and Rb–Sr mica ages in rocks of different ages (Barron 1966/1969; Priem et al. 1971; see Sect. 6.3), connected with the Grenvillian collision between the Laurentian and Amazonian Cratons (Kroonenberg 1982, 2019). The first K–Ar and Rb–Sr analyses on biotite from granites in the Colombian part of the Guiana Shield also gave ages around 1200 Ma (Pinson et al. 1962).

The year 1963 saw the breakthrough of geochronological methods in Guianan stratigraphy. The first bombshell was the paper *Age of the Roraima Formation, British Guiana* by Snelling (1963) from the Oxford Museum. Using the K–Ar method, he established the ages of a dolerite intruding in the Roraima Formation and of a hornfels at the contact of the dolerite with Roraima shale at about 1710 and 1735 Ma, respectively. McDougall et al. (1963) argued that the dolerite may even be as old as 2090 Ma. At the 6th Inter Guiana Geological Conference of 1963 in Belém do Pará, Brazil, Martin-Kaye (1966) showed first results with K–Ar and Rb–Sr mineral ages and Rb–Sr whole rock ages of a whole range of different rock units throughout British Guiana, obtained by Snelling at Oxford.

Shortly after, McConnell et al. (1964) presented a completely revised stratigraphy of British Guiana based on Snelling's radiometric ages in *Nature*. A specific paper on the geochronology of Guyana was presented at the 7th IGGC in Paramaribo in 1966 and published in 1969 (Snelling and McConnell 1969). Almost at the same time, Choubert (1964) published a series of mineral analyses from different rock types from French Guiana, which gave consistent grouping of Rb–Sr ages between 1832–1945 Ma and K–Ar ages from 1901–2053 Ma for biotites, muscovites and lepidolites, and Pb-Pb ages on zircon, columbotantalite, galena and ilmeno-rutile around 2220 Ma, with a few older ages around 2500 and 3770Ma. In Venezuela the first Rb–Sr mineral ages were obtained from the Guiana Shield by Shell, including on a biotite from a possibly pre-Imataca gneiss 2340 ± 55 Ma, on a biotite from Imataca itself 1540 ± 60 Ma, and a biotite from a granite 2000 ± 45 Ma (Short and Steenken 1962; Kalliokoski 1965).

At the University of São Paulo in Brazil geochronology also had already started in 1964 under the leadership of Umberto Cordani (Reynolds 2002), though initially not in the Guiana Shield but in southern Brazil (Amaral et al. 1966).

From 1966 onward also K–Ar and Rb–Sr radiometric ages became available from the basement of Suriname (Priem et al. 1966, 1967, 1968a,b, cf. Holtrop 1968), culminating in a review paper establishing a Rb–Sr isochron of the granitoid rocks 1874 ± 40 Ma (Priem et al. 1971; all Rb–Sr ages are recalculated with the internationally accepted Rb–Sr decay constant of $\lambda = 1.42 * 10^{-11} y^{-1}$). In 1973 he published the first age data on the volcanic ashes intercalated in the Roraima Formation at Tafelberg in Suriname, showing an Rb–Sr isochron age of 1655 ± 18 Ma (Priem et al. 1973).

A landmark paper was published in *Science* by Hurley et al. (1967) *Test of Continental Drift Comparison of Radiometric Ages*, a collaboration between the teams of the Massachusetts Institute of Technology and the University of São Paulo (USP). On the basis of about 150 new and published K–Ar and Rb–Sr age data Hurley et al. (1967) convincingly demonstrated that the four age provinces around 2700, 2000, 1000 and 600 Ma in northern South America closely matched similar age provinces in West-Africa, thus underscoring the validity of the continental drift paradigm (Fig. 3.1). As a counterpart to the 2000 Ma Eburnean Orogenic Cycle in West-Africa they called the coeval event in the Guiana Shield *Trans-Amazonian Orogenic Cycle*, a term that since then has obtained general acceptance.

The proliferation of radiometric data from the individual Guianian countries in the 1960ies resulted in an upsurge of correlation initiatives and new geochronology-based shield-wide geological maps in the 1970ies. While in the 7th IGGC in Paramaribo in 1966 the only paper on geochronology was the one by Snelling and McConnell, published in 1969 in *Geologie en Mijnbouw*, geochronology was an important subject at the 8th IGGC in 1970 in Georgetown, Guiana. Here Mc Connell and Williams (1970) presented the first geochronology-based map and correlation table of the Guiana Shield (Figs. 3.2 and 3.3).

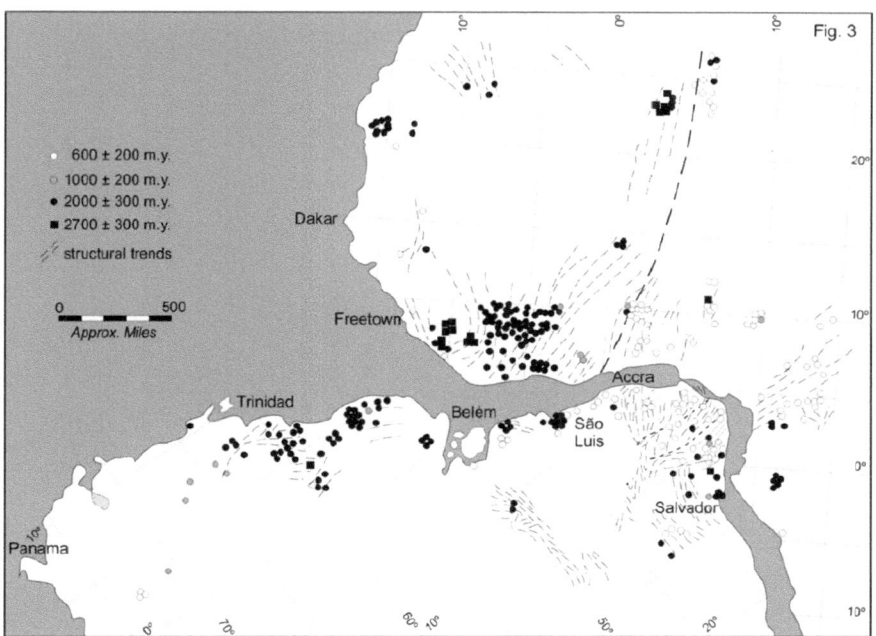

Fig. 3.1 Correspondence age provinces between South America and West Africa (after Hurley et al. 1967, redrawn by Cordani 2017)

In the accompanying paper, they subdivided the shield in three age groups. The Archean > 2500 Ma is represented by high-grade gneisses and granulites in the Imataca block in Venezuela, the Adampada-Falawatra (Bakhuis) area in western Suriname and the Île de Cayenne high-grade rocks. A geosynclinal series of Lower Proterozoic metasediments and metavolcanics younger than 2500 Ma but older than 2000–1800 Ma granites forms a mobile belt along the northern Atlantic coast, interrupted by the Takutu graben in Guyana. The western part of this belt in Venezuela and Guyana has a more eugeosynclinal character than the eastern part in Suriname and French Guiana. South of the eugeosynclinal belts there is a shelf facies with orthoquartzites and felsic volcanics (ignimbrites). The underlying Archean basement including the shelf facies cover has been reactivated and regranitized during the Trans-Amazonian orogeny. The third age group is represented by the tabular sandstones of the Roraima Formation, which are younger than the granites but older than basic intrusives dated at 1700 Ma. The authors stress the similarity between the geology of the Guiana Shield and that of West-Africa.

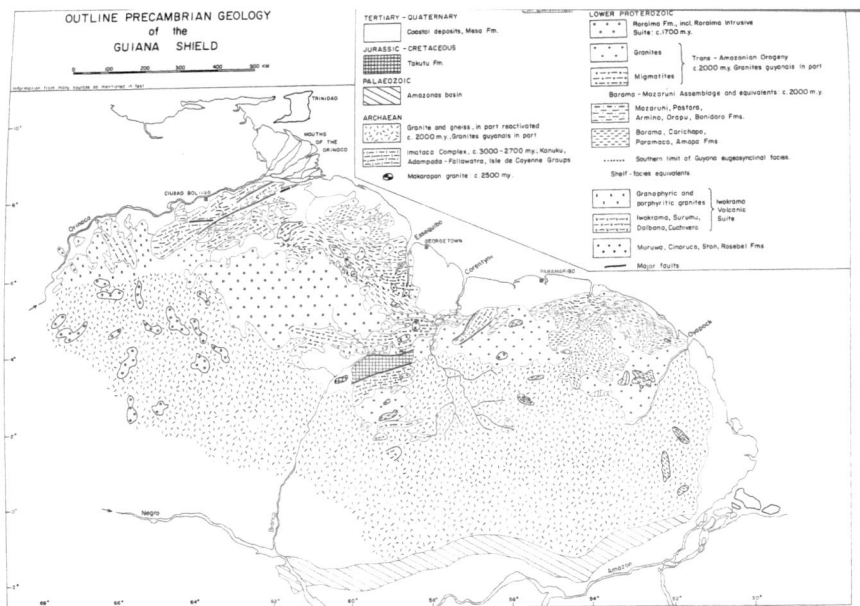

Fig. 3.2 First map of the Guiana Shield based on radiometric age data (McConnell and Williams 1970)

Boris Choubert was the first to publish a coloured geological map of the Guiana Shield (Fig. 3.4) in his book *Le Précambrien des Guyanes* (1974). It is a beautiful map with much detail, but his concept about the geotectonic evolution was still traditional: deposition in geosynclines, formed in grabens in an older Archean 'hyléen' basement, then a first mafic volcanic sequence, followed by an episode of folding and 'granitization' (granite guyanais of 2200–2300 Ma), then deposition of flysch-type sediments and a second wave of folding and granitization (granite caraïbe, 1872–2100 Ma). It appeared difficult to correlate the events in French Guiana with other parts of the shield, as the French colony is almost exclusively underlain by granite-greenstone terrain.

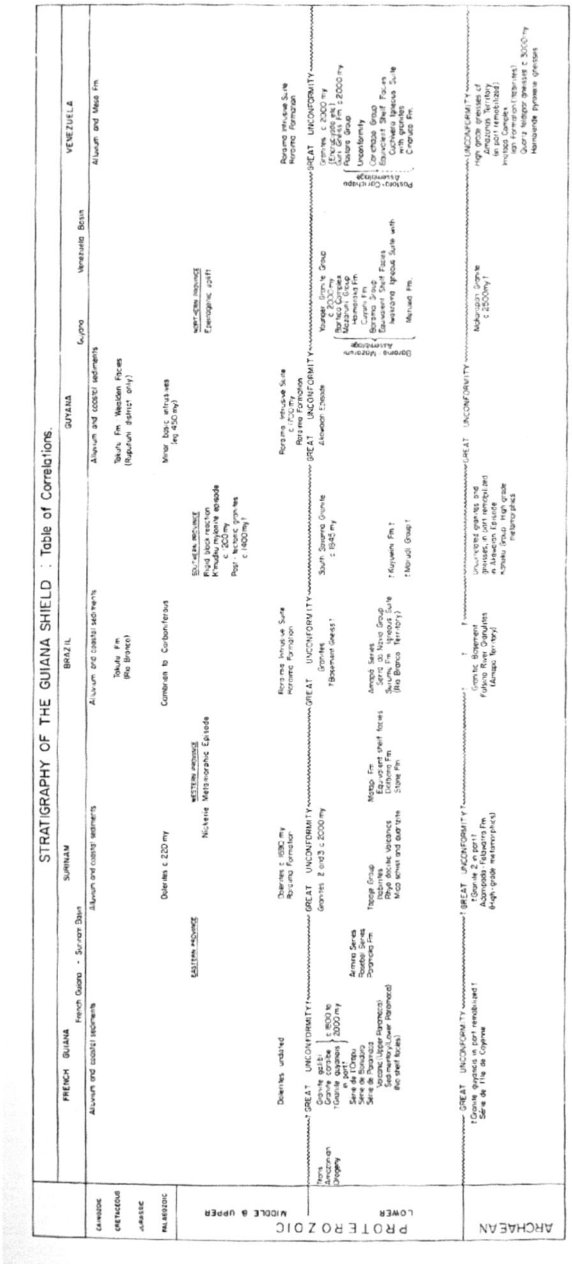

Fig. 3.3 The first lithostratigraphic scheme of the Guiana Shield based on radiometric data (McConnell and Williams 1970)

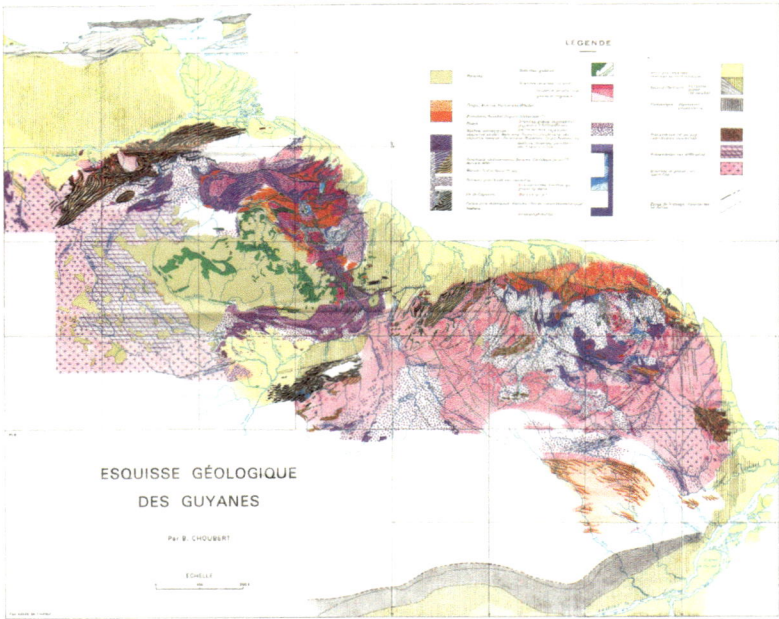

Fig. 3.4 The first coloured geological map of the (northern) Guiana Shield by Choubert (1974)

References

Amaral G, Cordani UG, Kawashita K, Reynolds JH (1966) Potassium-argon data of basaltic rocks from southern Brazil. Geochim Cosmochim Acta 80:169–189

Barron CN (1969) Notes on the stratigraphy of Guyana. In: Proceedings of the seventh Guiana geological conference, Paramaribo, 1966. Records Geol Surv Guyana 6(II):1–28

Choubert B (1964) Ages absolus de la Guyane Française, (Précambrien guyanais) CR Acad Sci Paris 258:631–634

Choubert B (1974) Le Précambrien des Guyanes. Mémoires BRGM 81, p 213

Cordani UG (2017) História geológica do Craton Amazônico Anais do 15° Simpósio de geologia da Amazonia, Belém:11–16

Holtrop JF (1968) The stratigraphy and age of the Precambrian rocks of Surinam. Geol Soc Am Bull 79:501–508

Hurley PM, de Almeida FFM, Melcher GC, Cordani UG, Rand JR, Kawashita K, Vandoros P, Pinson WH, Fairbairn HW (1967) Test of continental drift by comparison of radiometric ages. Science 157:495–500

Kalliokoski J (1965) Geology of north-central Guayana Shield, Venezuela. Geol Soc Am Bull 76:1027–1050

Kroonenberg SB (1982) A Grenvillian granulite belt in the Colombian Andes and its relation to the Guiana Shield. Geol Mijnbouw 61(325):333

Kroonenberg SB (2019) The proterozoic basement of the western Guiana Shield and the northern andes. In: Cediel F, Shaw RP (eds) Geology and tectonics of northwestern South America, Front Earth Sci. Springer, pp 115–192

Martin-Kaye PHA (1966) Progress by the geological survey of British Guiana. In: Anais da VI conferência geológica das Guianas, 1963, Belém, pp 113–121

McConnell RB, Dixon CG (1961) Presentation of the revised geological map of British Guiana. In: Proceedings of the fifth inter-Guiana geological conference, Georgetown, British Guiana, pp 17–30

McConnell RB, Cannon RT, Williams E, Snelling NJ (1964) A new interpretation of the geology of British Guiana. Nature 204:115–118

McConnell RB, Williams E (1970). Distribution and provisional correlation of the Precambrian of the Guiana Shield. In: Proceedings of the eighth Guiana geological conference, Georgetown, Guyana, vol I, pp 3–23

McDougall I, Compston W, Hawkes DD (1963) Leakage of radiogenic argon and strontium from minerals in Proterozoic dolerites from British Guiana. Nature 198:564–567

Pinson WH, Hurley PM, Mencher E, Fairbairn HW (1962) K-Ar and Rb-Sr ages of biotites from Colombia, South America. Geol Soc Am Bull 73:807–910

Priem HNA, Boelrijk NAIM, Verschure RH, Hebeda EH (1966) Isotopic age determinations on Surinam rocks. Geol Mijnbouw 45:16–19

Priem HNA, Boelrijk NAIM, Verschure RH, Hebeda EH (1967) Isotope age determinations on Surinam rocks, 2. Geol Mijnbouw 46:482–486

Priem HNA, Hebeda EH, Boelrijk NAIM, Verschure RH (1968a) Isotope age determinations on Surinam rocks, 3. Proterozoic and permo-triassic basalt magmatism in the Guiana Shield. Geol Mijnbouw 47:17–20

Priem HNA, Hebeda EH, Boelrijk VRH, Verdurmen EAT (1968b) Isotopic age determination on Surinam rocks 4. Ages of basement rocks in North-Western Surinam and of the Roraima Tuff at Tafelberg. Geol Mijnbouw 47:191–196

Priem HNA, Boelrijk NAIM, Hebeda EH, Verdurmen EAT, Verschure RH (1971) Isotopic ages of the trans-amazonian acidic magmatism and the Nickerie episode in the Precambrian basement of Surinam, South America. Geol Soc Am Bull 82:1667–1680

Priem HNA, Boelrijk NAIM, Hebeda EH, Verdumen EAT, Verschure RH (1973) Age of the Precambrian Roraima formation in northeastern South America: evidence from isotopic dating of Roraima pyroclastic volcanic rocks in Suriname. Geol Soc Am Bull 84:1677–1684

Reynolds JH (2002) Origin of the Center for Geochronological Research at São Paulo. Revista do Instituto de Geociências—USP 2–1–8

Short KC, Steenken WF (1962) A reconnaissance of the Guayana Shield from Guasipati to the Rio Aro, Venezuela. Asociacion Venezolana de Geologia, Mineria y Petroleo (AVGMP), Boletín Informativo 5(7):189–217

Snelling NJ (1963) Age of the Roraima formation, British Guiana. Nature 198:1079–1080

Snelling NJ, McConnell RB (1969) The geochronology of Guyana. Geol Mijnbouw 48:201–213

Chapter 4
Precambrian Fixism or Plate Tectonics? Venezuela Leads the Way

Abstract The availability of radiometric data and the upsurge of the plate tectonics concept in the 1960ies and 1970ies led to the first ideas about the tectonic history of the Guiana Shield as a whole. Some authors maintained that Precambrian tectonics were different than at present, citing authors working in the Archean of southern Africa, others directly transposed Phanerozoic plate tectonic processes to the Precambrian.

The paradigm shift to plate tectonics in the 1960ies led to a surge of new ideas, which found their way also in the Guiana Shield. Aerogeophysical surveys over many parts of the shield revealed unsuspected structural detail. Furthermore, after the experiments of Bowen and Tuttle (1950) and later Winkler and Von Platen (1961) it was obvious that granites usually form from melts, not by solid-state metasomatism. It heralded the end of the 'granitization' concept which up to the 1970ies marred many descriptions of 'granites' which we now would classify as migmatites.

A major breakthrough in the context of plate tectonics was the recognition by McConnell (1969) that the NE-SW stretching Jurassic-Cretaceous Takutu rift valley in Guyana was a corollary of the opening of the North Atlantic Ocean. The rift valley separates the Guiana Shield in an eastern and a westen part, that are difficult to correlate with each other (see Chap. 9). But whether plate tectonics already played a role in the Precambrian is not discussed in his papers.

In Venezuela Cecilia Martín-Bellizzia (1972) presented an extensive paper on Paleotectonics of the Guiana Shield at the 9th Inter Guiana Geological Conference in Ciudad Guayana, Venezuela. Her views about the earliest history of the shield appear to have been inspired by the 'classical' fixists concepts of Anhaeusser et al. (1969) and other authors in southern Africa (Fig. 4.1) in which oval older cratons are surrounded by younger 'mobile belts' of different ages. She divides the Guiana Shield, with its overall oval shape in itself, in two smaller more or less oval cratonic parts, separated by the Takutu-Bakhuis high-grade belt as defined by McConnell and Williams (1970) (Figs. 4.2 and 4.3). The northern oval occupies most of eastern Venezuela and Guyana, the southern one Suriname, French Guiana and the whole still ill-defined Brazilian part of the Guiana Shield. According to her,

S. Kroonenberg, *The Changing Framework of the Guiana Shield*, SpringerBriefs in Earth System Sciences, https://doi.org/10.1007/978-3-031-86334-9_4

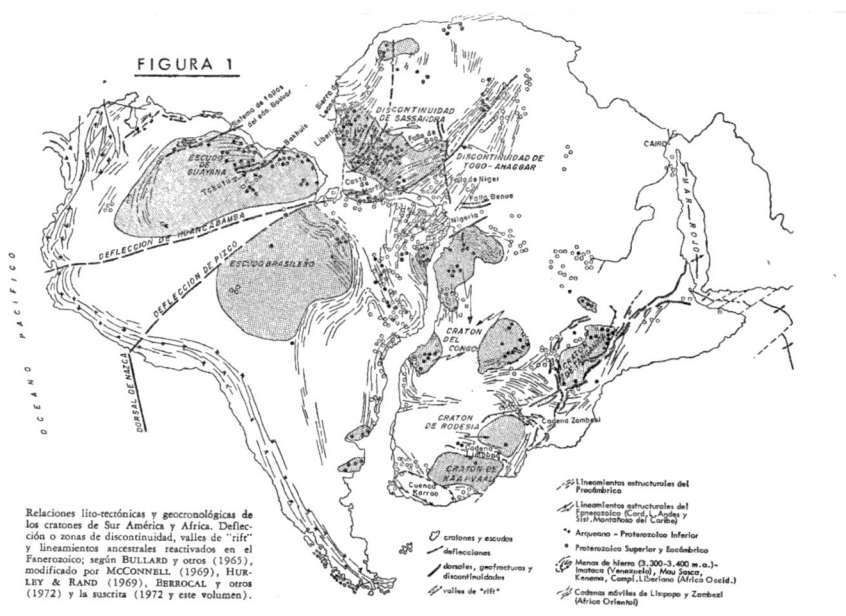

Fig. 4.1 Geofractures between oval cratons in South America and Africa (Martín-Bellizzia 1972)

both ovals are associated with deep dislocations or 'geofractures' that seem to correspond to mantle lineaments from the Early Archean. The ovals are surrounded by
discontinuous Archean granulitic belts (Fig. 4.2). The deep geofractures gave rise
to horsts and grabens that were reactivated during the four main tectonomagmatic
episodes Guriense (3400–3000 Ma), Aroense (2750–2650 Ma), Trans-Amazonian
(2000–1800 Ma) and Orinoquensis-K'Mudku-Nickerie (1200–850 Ma) (Fig. 4.4).
    Within the cratonic ovals she distinguishes amphibolitic and greenstone belts.
The amphibolite belts are characterised by amphibolite-facies gneisses, metacherts,
basaltic metalavas, amphibolites and (ultra)mafic intrusives. They differ from ophi-
olites in modern mountain belts, and are thought to represent the initial phase of the
development of the rifts in the two cratonic ovals of the shield. Their ages around
2300–2000 Ma may be masking an older event around 2700 Ma as in Imataca.
The greenstone belts encompass greenschist-facies phyllites with black carbonatic,
manganiferous and ferruginous intercalations, arkoses, dacitic metalavas and associated pyroclastics, and locally also mafic lavas. She states that, in contrast to most
scientists who prefer an origin in oceanic spreading zones, she believes that these
belts are the result of intracontinental rifts in a thinned Archean crust.

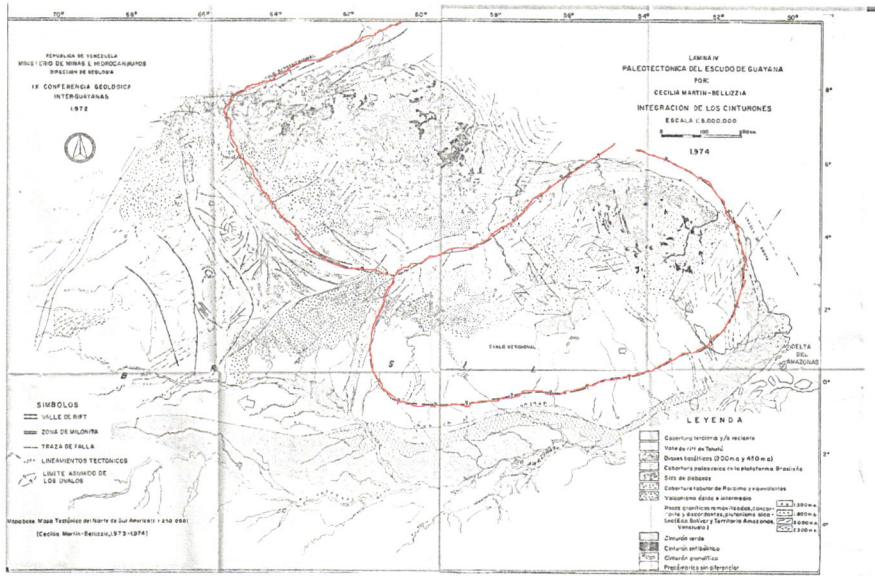

Fig. 4.2 The two oval cratonic units within the Guiana Shield (modified after Martín-Bellizzia 1972)

Fig. 4.3 Tectonic wall map of Northern South America 1:2.5 000 000 (Martín-Bellizzia 1978). The geological units and their delimitations in the Guiana Shield are largely the same as in Fig. 4.2

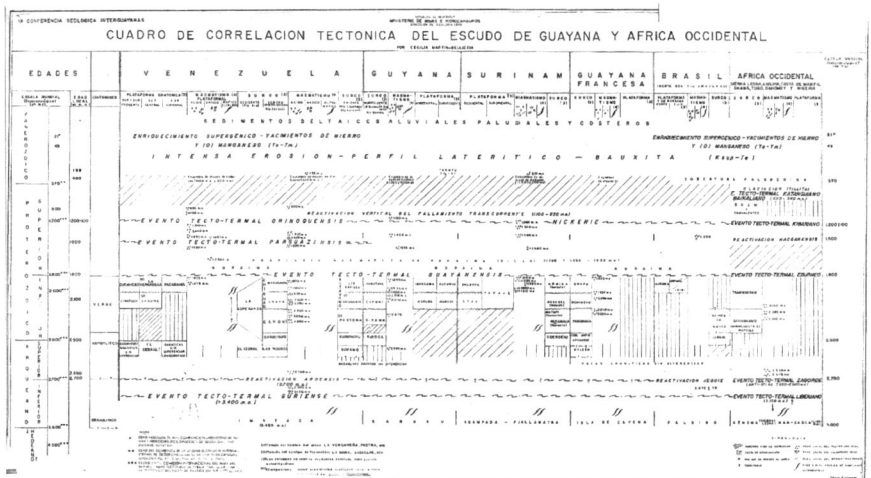

Fig. 4.4 Correlation scheme of the Guiana Shield with West-Africa (Martín-Bellizzia 1972)

Three phases of granitoid magmatism are distinguished: a catazonal one associated with banded and migmatitic gneisses in the granulite belts, syntectonic circular or ellipsoidal bodies of granodioritic (sodic) series associated with the amphibolite and greenstone belts, and a later mainly post-tectonic porphyric potassic series. Granitoid magmatism is manifest in six periodes of mobilization and intrusion from 3400 to 850 Ma. Martín-Bellizzia embraces the plate-tectonic concept of the Mesozoic separation of South America and Africa along the Mid-Atlantic Ridge, but sea-floor spreading and orogeny through subduction processes in the Precambrian is rejected.

At present we would consider the amphibolite belts, the greenstone belts and the associated granites essentially as characteristic components of the large Paleoproterozoic Maronian greenstone belt along the North-Atlantic coast, the equivalent of the Birimian in West-Africa.

Ovals were in vogue in those times: the inventive Dutch geologist Han Kloosterman presented a paper at the 2nd Latin-American Geological Conference in Caracas in 1973 with the title *Giant Ring Volcanoes on the Guiana Shield*. On his map (Fig. 4.5) he distinguished three huge ring structures, the Roraima volcano of 500 × 900 km, the Suriname volcano of 300 × 350 km and the less well known Amazon volcano. The Roraima and Tafelberg sandstone formations would then represent the caldera filling of the first two mentioned. They coincide to some extent with the ovals of Martín-Bellizzia (1972), being bordered by granulite belts as Imataca and Bakhuis, respectively, and having the 'geosynclinal' sequences of the greenstone belts to the north. These giant calderas would have been responsable for the extensive felsic volcanic and plutonic magmatism in large parts of the central Guiana Shield. His views, though no longer tenable, foreshadow the later recognition of the felsic magmatism as being produced by the Orocaima and Uatumã mantle plumes (Reis et al. 2021).

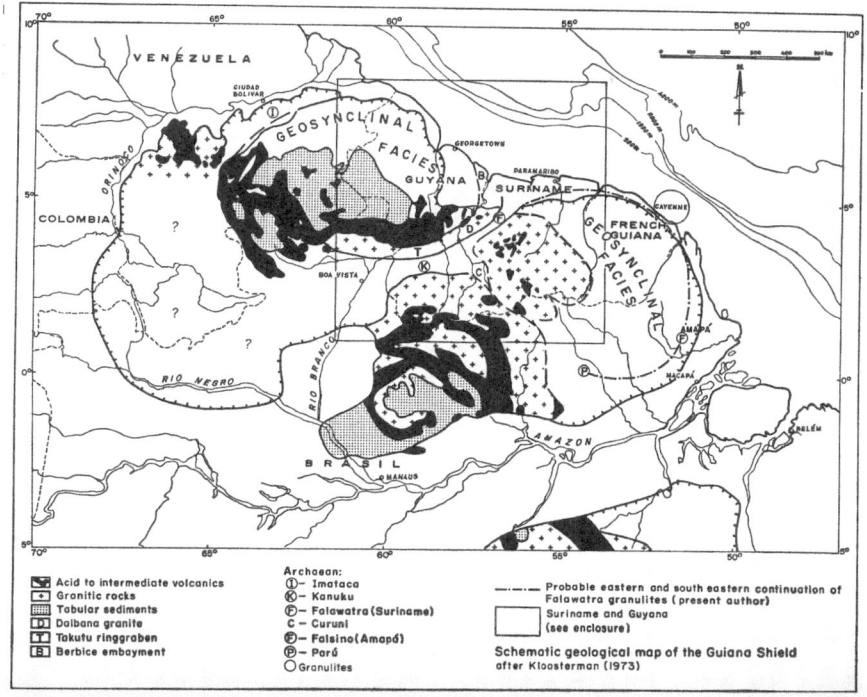

Fig. 4.5 Giant calderas (Kloosterman 1976)

Oval structures were also in vogue on smaller scales. Within the granulite belts of Suriname and Guyana, the Surinamese geologist Henk Dahlberg recognized the Bakhuis, Corantijn and Kudiditau oval domal structures (Dahlberg 1975), inspired by the work of the Russian geologist Lazarus Salop (1971) who described similar structures in the Aldan shield in Siberia. And Choubert (1965) published an idealised picture of the diapiric elliptical TTG intrusions in the greenstone belt of French Guiana, a concept eagerly adopted also by Cecilia Martín Bellizzia in her 1972 paper.

At the same conference in 1973 in Caracas, a colleague of Martín-Bellizzia at the Ministry of Mines and Hydrocarbons, Vicente Mendoza, presented his own views on the tectonic evolution of the Guiana Shield. His well-written analyses speak of orogenies and subduction processes. There must have been vivid discussions at the Ministry between Mendoza en the Bellizzia's. He started his paper with an overview of all

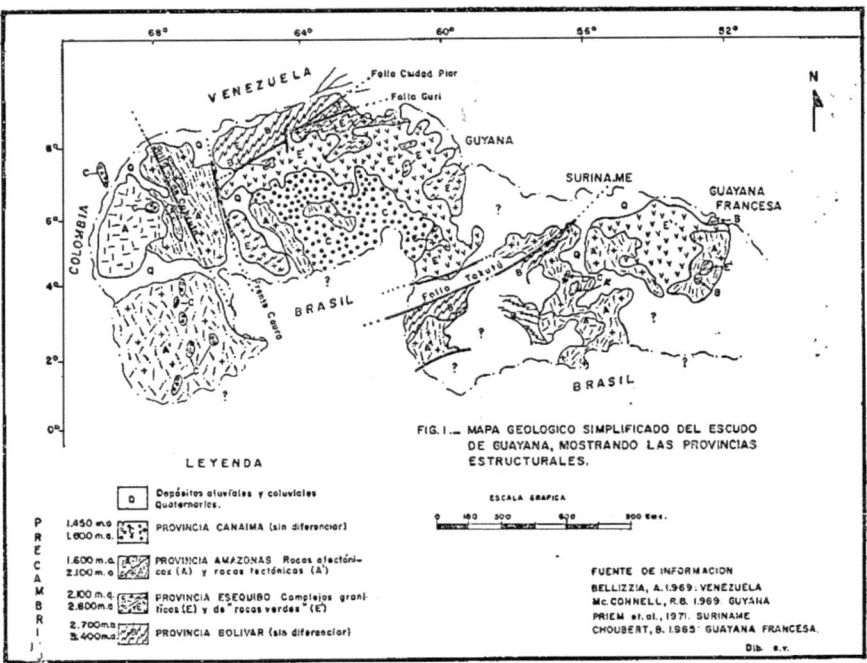

Fig. 4.6 The four 'provinces' represent from old to young the granulite belts, the greenstone belts, the felsic magmatism and Roraima as identified by Vicente Mendoza (1977)

radiometric ages obtained in the shield thusfar. He subdivided the Shield in four structural provinces, and attributes their origin to four orogenic events, respectively the Gurian (3600–2700 Ma), PreTrans-Amazonian (2600–2100 Ma), Trans-Amazonian (2000–1700 Ma) and Orinoquean orogenies (1200–800 Ma) (Fig. 4.6), thus largely maintaining the fourfold grouping of ages of Martín-Bellizzia. He noted the absence of a primordial oceanic crust, as the oldest structural province Bolívar (including Imataca, Kanuku, Bakhuis, Île de Cayenne, the granulite belt of Martín-Bellizzia) consists of high-grade metamorphic metasediments without much mafic elements. Subduction leading to the formation of an island arc, sedimentation, calcalkaline volcanism and plutonism are the main processes in its origin, followed by continental arc orogeny and intense folding into isoclinal structures (Fig. 4.7). The second province called Essequibo, encompassing the main greenstone belts of the shield, started first with ocean spreading, followed by a second island arc and finally orogenic accretion to the older Bolívar province. This reconstruction is already much closer to modern concepts of the history of the development of the Guiana Shield. Recently Mendoza published an extensive book on the geology of Venezuela (Mendoza 2012).

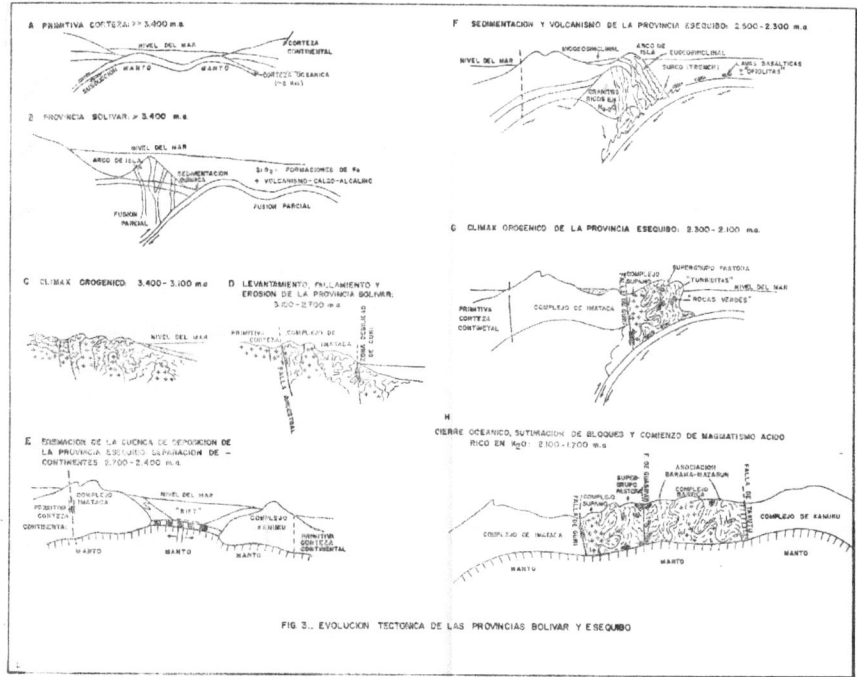

Fig. 4.7 The first modern reconstruction of the history of the Guiana Shield (granulite and greenstone belts) (Mendoza 1977)

However, as in almost all previous evolutionary schemes, the southern, Brazilian part of the shield is largely neglected by the Venezuelans. At the other side of the frontier, the Brazilian geologist Elton Suszczynski (1970) recognised two orogenic phases, one in Amapá for which he cited one Rb–Sr age of 2000 Ma obtained by Umberto Cordani in São Paulo, and one in the southern Guiana Shield, the North-Amazonian orogeny, still subdivided in a Parú and a Rio Negro phase (Fig. 4.8). He was aware of what had been done in the other Guianan countries but refrained from making correlations.

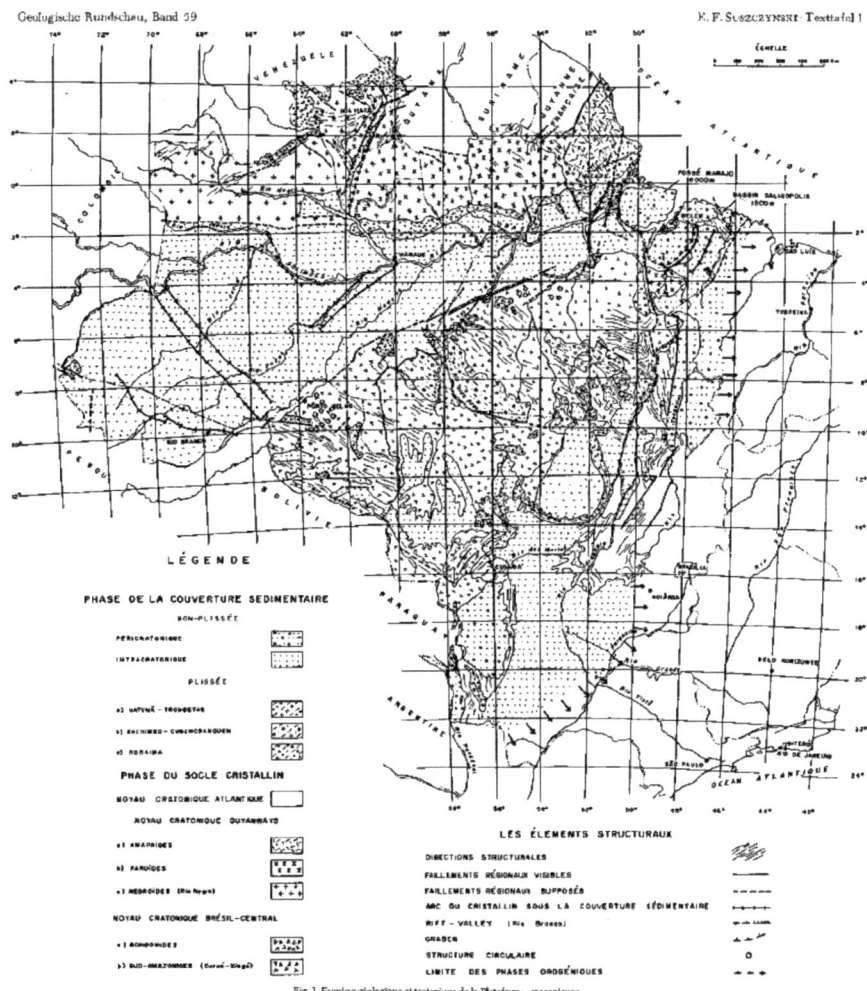

Fig. 4.8 Suszczynski's map of the Amazonian Craton: only Brazilian part of the Guiana Shield depicted

References

Anhaeusser CR, Mason R, Viljoen M, Viljoen RP (1969) A reappraisal of some aspects of Precambrian shield geology. Geol Soc Amer Bull 80:2175–2200

Bowen NL, Tuttle OF (1950) The system $NaAlSi_3O_8$-$KAlSi_3O_8$-H_2O. J Geol 58:489–511

Choubert B (1965) État actuel de nos connaissances sur la géologie de la Guyane française. Bulletin de la Société Géologique de France; S7–VII(1):129–135

Dahlberg EH (1975) Lithostratigraphical correlation of granulite-facies rocks of the Guiana Shield. Med Geol Mijnb Dienst Suriname 23:26–33. Also published (1976) in: Memorias 2do Congreso Latinoamericano de Geología, Caracas, 1973, Boletín de Geología, Publ Esp 7:665–673

Kloosterman JB (1976) Giant ring volcanoes on the Guiana Shield. Memoria Segundo Congreso Latinoamericano de Geología, Caracas, 1973. Boletín de Geología, Publicación especial 7:713–722

Martín-Bellizzia C (1972) Paleotectónica del Escudo de Guayana. Memoria de la 9na conferencia geológica Inter-Guayanas. Boletín de Geología (Caracas), Publicación especial No. 6:251-304

Martín-Bellizzia C (1978). Mapa tectónico Norte de America del Sur, 1:2.5 000 000, Ministerio de Minas y Energía, Caracas

McConnell RB (1969) Fundamental fault zones in the Guiana and West African shields in relation to presumed axes of Atlantic spreading. Geol Soc Am Bull 80:1775–1782

McConnell RB, Williams E (1970). Distribution and provisional correlation of the Precambrian of the Guiana Shield. In: Proceedings of the eighth Guiana geological conference, Georgetown, Guyana, vol I, pp 3–23

Mendoza V (1977) Evolución tectónica del Escudo de Guayana. Segundo Congreso Latinoamericano de Geologia. Boletín Geológico Caracas Publicación Especial 7(III):2237–2270

Mendoza V (2012) Geología de Venezuela. Gran Colombia Gold Comp., Bogotá, Colombia, p 362

Reis NJ, Teixeira W, D'Agrella-Filho MS, Bettencourt JS, Ernst RE, Goulart LEA (2021) Large igneous provinces of the Amazonian Craton and their metallogenic potential in Proterozoic times. In: Srivastava, R. K., Ernst, R.E., Buchan, K. L. and de Kock, M. (eds) 2022. Large Igneous Provinces and their plumbing systems. Geological Society, London, Special Publications 518:493–529

Salop L (1971) Two types of precambrian structures: gneiss folded ovals and gneiss domes. Bull Moscow Soc Nat 4:5–30. Geological series, Transl in: IntGeol Rev 14:1209–1228. https://doi.org/10.1080/00206817209475823

Suszczynski EF (1970) La Géologie et la Tectonique de la Plateforme Amazonienne. Geol Rundsch 59:1232–1253

Winkler HGF, von Platen H (1961) Experimentelle Gesteinsmetamorphose—IV: bildung anatektischer Schmelzen aus metamorphisierten Grauwacken. Geochim Cosmochim Acta 24:48

Chapter 5
The Radar Mapping Era: A Brazilian Role Model

Abstract Remote sensing imagery, especially side-looking radar and LANDSAT, enabled great advances in geological mapping of the rainforest-clad Guiana Shield. Especially in Brazil, but also in Colombia and Venezuela spectacular improvements were obtained in the delimitation of geological units and structural features.

The 10th Inter Guiana Geological Conference in Belém do Pará in 1975 was an enormous revelation. It marked the presentation of the Projeto RadamBrasil, an ambitious inventory of the geology, geomorphology, soils, vegetation and land use potential of the whole Amazonian territory of Brazil at 1:1 000 000 scale, using radar imagery and field surveys. At that time I was employed at the Geological Mining Service of Suriname, and it was my first international congress. It was hugely impressive. One of the conference rooms was covered from wall to wall with mosaics from radar images of the nine sheets already published. For each sheet a comprehensive text volume had been published with all data on natural resources, including geological maps, stratigraphical tables and each lithological unit was extensively described with characteristic radar images, structural trends, petrography, geochemistry, radiometric ages and mineral occurrences. Four volumes covered parts of the Guiana Shield, e.g. Oliveira et al. (1975), and more were to follow after that year. It marked a revolution in the way the natural resources of the densely forested Amazonian Craton was documented.

Issler (1975) opened the proceedings of the conference with a review and a detailed but irreproducible map of the Brazilian part of the Guiana Shield, with also reference to data from the other Guianian countries (Fig. 5.1). I will discuss these findings somewhat extensively as they will play an important role in discussions later in this book.

The oldest orogenic cycle, called Gurian as in Venezuela, is evidenced in the high-grade Complexo Guianense gneisses and granulites in Amapá with ages between 2600 and 2400 Ma, the first evidence of the presence of an Archean block in that area. A single K–Ar biotite age of 2531 ± 12 Ma from the Anauá area in Roraima province was later considered doubtful (Oliveira et al. 1975). Other areas included in the Gurian orogeny gave Rb–Sr ages around 1800 Ma, interpreted as being due to

© The Author(s), under exclusive license to Springer Nature Switzerland AG 2025   41
S. Kroonenberg, *The Changing Framework of the Guiana Shield*, SpringerBriefs in Earth System Sciences, https://doi.org/10.1007/978-3-031-86334-9_5

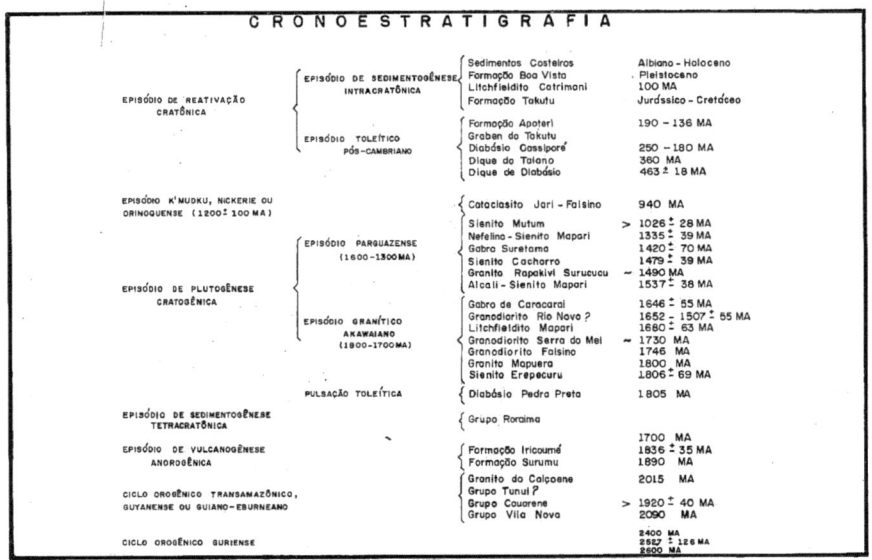

Fig. 5.1 Chronostratigraphy of the Brazilian part of the Guiana Shield, Issler (1975)

a later Trans-Amazonian metasomatic overprint over an older massif. We are here still in the granitization mode.

The Gurian event was followed by the Trans-Amazonian Orogeny between 2000 and 1920 Ma, and affected the Vila Nova Group, consisting of amphibolite-facies metamorphosed mafic volcanics and chemical sediments in the Amapá area, giving a Rb–Sr isochron of 2090 Ma. This zone forms part of the extensive greenstone belt in the northern Guianan countries. Schists and quartzites from the Cauarane Group in Roraima province dated at about 1920 ± 40 Ma are also included in this unit by Issler. The Roraima sandstones and conglomerates cover these units unconformably, and are intruded by dolerites that gave K–Ar ages around 1600 Ma, in the same range as the volcanic ash dated by Priem et al. (1973), see discussion above.

Then follows an anorogenic event with extensive felsic to intermediate volcanics, including pyroclastics and ignimbrites, collectively called Uatumã Group, with ages between 1890 and 1700 Ma. Later they will appear to represent two separate events, the earlier Orocaima and later Uatumã events (see Chaps. 6 and 8). The relation between these volcanics and those in the Roraima Formation is still unclear at that time.

More or less in the same interval granitic bodies intruded during the so called Akawaian Granitic Episode between 1800 and 1700 Ma. The term Akawaian, an unfortunate but persistent misnomer, was apparently borrowed from Williams et al. (1967), who used it for the (Rhyacian) TTG bodies intruding into the Trans-Amazonian greenstone belts of Guyana (Snelling and McConnell 1969). Gibbs (1980, p. 245) in his greenstone thesis regards 'Akawaian' as synonymous to 'Trans-Amazonian', just as Berrangé (1977) in his extensive review of the geology of southern Guyana.

A still younger magmatic event produced rapakivi and other alkaline granites between 1600 and 1300 Ma, called Parguaza event after the coeval Parguaza rapakivi granite in Venezuela. The last tectonic event is the already mentioned mylonitic and tectonothermal event around 1200 Ma, called Orinoquian in Venezuela, K'Mudku in Guyana, Nickerie in Suriname and Jari-Falsino in Brazil (Issler 1975).

In general, Issler's scheme shows much resemblance with findings from the northern part of the Guiana Shield in Venezuela, Guyana, Suriname and French Guiana, as discussed above. But there is no reference to the nature of the geotectonic processes that gave rise to these developments.

At the same time there seems to have been a lively discussion in São Paulo on the tectonic significance of the new data. In 1974, one year before the conference, the Brazilian geologist Gilberto Amaral produced his excellent *livre-docência* thesis *Geologia Pré-Cambriana da região Amazônica*. The *livre-docência* in Brazil is a higher level degree than the PhD-doctorate, required to become a professor. Amaral had been an assistent at USP in the 1960ies with Umberto Cordani (Reynolds 2002), but he produced his thesis while being employed by INPE, Instituto Nacional de Pesquisas Espaciais, the Brazilian NASA, to make a geological map of the Amazonian area on the basis of radar and LANDSAT imagery. Cordani is not even acknowledged in his thesis. The thesis presents the best synthesis for that time I have read so far of the Amazonian Craton, or Amazonian Platform as he calls it, following Suszczynski (1970). He carefully cites almost all papers by previous scientists from all Guianian and Amazonian countries that had been published so far. He also elaborated a large number of geochronological data from the whole area, partly also dated by himself after his fieldwork in the Roraima province. In the sequence of orogenic events he accepts the subdivision into Gurian (3400–3000 Ma), Guianian (2700–2500 Ma) and Trans-Amazonian (2000–1800 Ma), following most authors discussed above. The map in his thesis is a different one than the map produced by Radambrasil (Issler 1975). For some reason his thesis never appeared in print, and the details of his interpretations would only become internationally known ten years later, as we will see.

For practical purposes he subdivided the area on the basis of lithological associations in three NNW-SSE zones, stretching from the southern Guiana Shield across the Amazon basin into the adjacent Central Brazilian Shield: Eastern, Central and Western Amazonia (Fig. 5.2).

The northern part of the Eastern Amazonia Province in Amapá in the Guiana Shield consists of a basement of gneisses and granulites, followed by the Vila Nova Group exposed in the Serra do Navio manganese mine along the Amaparí river, from which one sample gave an Rb–Sr whole-rock age of 2619 Ma and five others 2044 to 2329 Ma. K–Ar mineral ages from other Amapá sites were 2084–1761 Ma and Phanerozoic ages. The last development was the intrusion of granites. In the southern part of the Eastern Amazonian province in the Guaporé Shield, south of the Amazon basin, the huge Carajás iron, copper and gold deposits were discovered in 1967, at that time generally also giving ages around 1900–2000 Ma, but later established as mainly Archean.

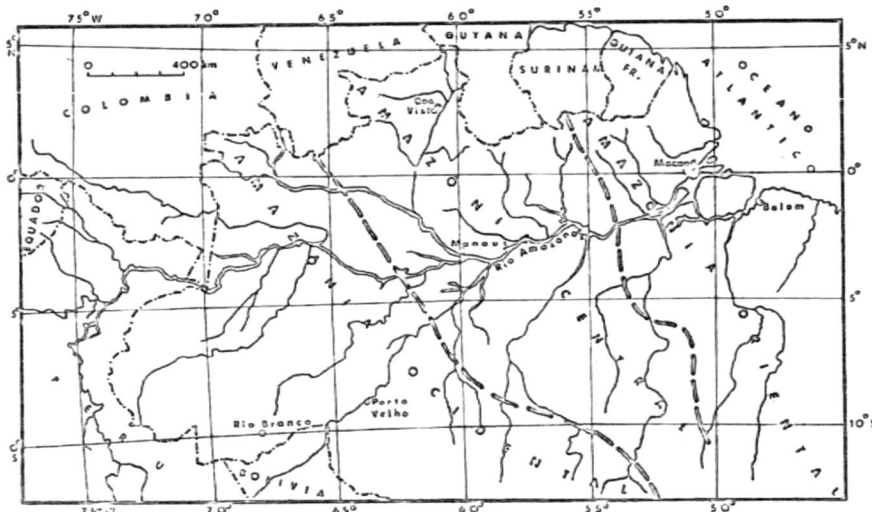

Fig. 5.2 Subdivision of the Brazilian part of the Amazonian Craton in an Eastern, Central and Western province (Amaral 1974)

In the northern part of the Central Amazonian Province Amaral distinguishes at the base two complexes: the Anauá-Uraricoera complex containing high-grade metamorphic gneisses, schists and amphibolites, and the Mucajaí complex, also granulites and charnockites. On top of these, Surumu felsic volcanics and associated Uailán granitoid intrusions are developed. The Surumu rocks give an Rb–Sr isochron age of 1635 ± 30 Ma, in the same range as Priem's ages from the Roraima tuffs. K–Ar mineral ages of the Surumu volcanics are around 1175 Ma, representing the K'Mudku-Nickerie mylonitic and thermal episode defined before. Roraima sandstone cover these series unconformably. In the southern part of the Central Amazonian Province Uatumã felsic volcanics were correlated with the Surumu in the northern part.

From the northern part of the Western Amazonian Province few data were yet available at that time. There is a basement with gneisses and granites, followed by metavolcanics, possibly related to the Surumu in the Central part, and by a sedimentary sequence, among others in the Tunuí and Taraira areas, as well as younger granitoid intrusions. There is only one Rb–Sr age of 1532 Ma, while K–Ar mineral ages range between 1524 and 1315 Ma. The southern part of this western province in the Guaporé Shield, coinciding with Rondônia state, well-known because of its tin-bearing granites, shows in general still much younger rocks, but it will not be discussed in the present paper, being far away from the Guiana Shield.

In a later chapter of his thesis he comprehensively compares the geology of the Amazonian Craton with data from the surrounding countries, and draws correlations between the greenstone belt formations in French Guiana, Suriname, Guyana and Venezuela with the Vila Nova Group in Amapá, between the high-grade belts in

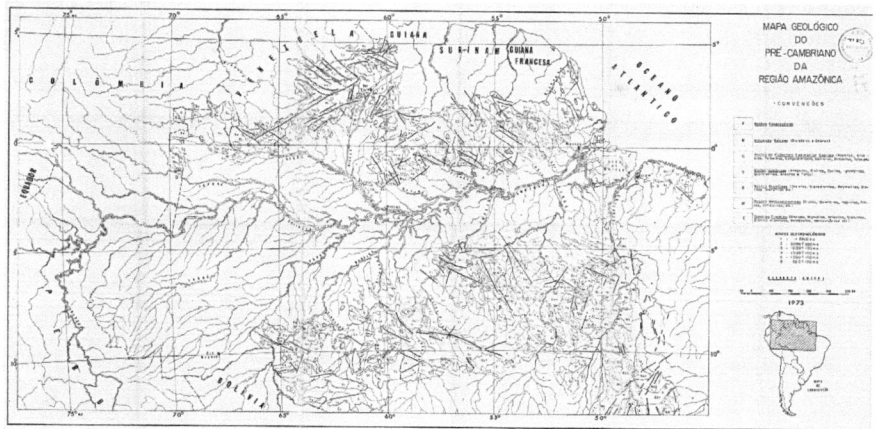

Fig. 5.3 Detailed geological map of the Brazilian part for the Amazonian Craton (Amaral 1974)

Venezuela, Guyana and Suriname and the Mucajaí complex, and between the felsic volcanics of Iwokrama with the Surumu volcanics in Brazil (Fig. 5.3).

While noting that the Imataca and Carajás complexes are the oldest in the Amazonian Craton, his geotectonic reconstruction starts with the notion that at the end of the Guianian event (around 2600 Ma), the whole northeastern portion of the craton was already stabilised and surrounded by the Trans-Amazonian geosynclinal basins we now consider as greenstone belts. At the end of the 'Middle Precambrian' they were folded and metamorphosed and underwent extensive magmatism. After a period of consolidation sedimentary formations including the Muruwa (Guyana) and Cinaruco (Venezuela) sandstone platforms were deposited, followed by the widespread felsic volcanism of the Surumu and related formations. Deposition of the Roraima Formation and related deposits elsewhere in the craton followed.

Gilberto Amaral's earlier views on the Amazonian Craton only appeared in print in 1984 in a book *O Précambriano do Brasil* edited by the nestor of geology at USP, Fernando Flávio Marques de Almeida and Yociteru Hasui. The editors of the book did not cite Cordani's 1979 paper (Chap. 6) in their introduction, but they gave Gilberto Amaral the opportunity to publish a paper on the Provincia Tapajós and the Provincia Rio Negro, the Brazilian parts of the Guaporé Shield and the Guiana Shield, respectively (Amaral 1984). Amaral stresses the importance of the use of remote sensing imagery in Amazonian terrains, which produce data that never can be obtained though fieldwork alone. He also warns against putting too much confidence on geochronology, because it indeed produces a large amount of relevant data, but often of doubtful interpretation. The values obtained are being interpreted according to the conviction of each author, he states, often in stark contradiction with stratigraphic data.

Amaral (1984) maintains the divisions of the Amazonian craton of his 1974 thesis in three NW–SE stretching zones, but subdivides each of them in separate

Fig. 5.4 Subdivision of the
Brazilian part of the
Amazonian Craton in
provinces and subprovinces
(Amaral 1984)

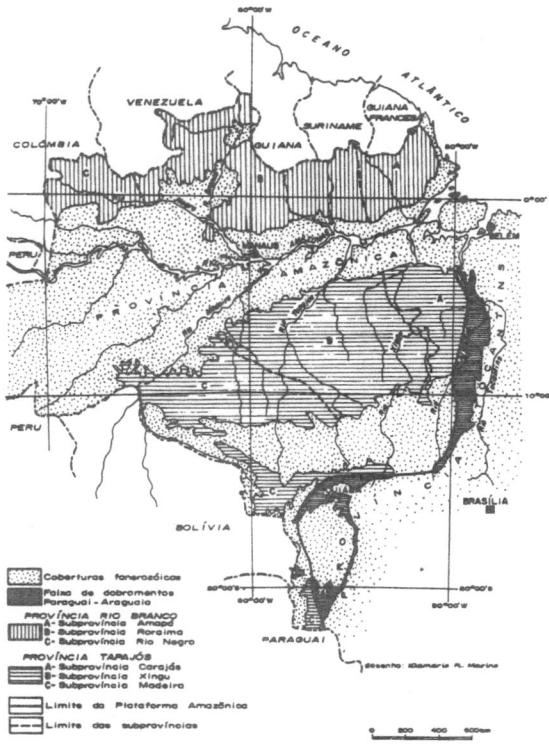

Figura 2-1 Caracterização e subdivisão das províncias Rio
Branco e Tapajós.

subprovinces: in the Guiana Shield from NE to SW Amapá, Roraima and Rio Negro
subprovinces, and in the Guaporé Shield south of the Amazon River Carajás, Xingu
and Madeira (Fig. 5.4). The latter will not be not discussed here. In this way there
is no need to discuss correlations across the Amazon basin. The stratigraphy and
geochronology of each subprovince is updated with respect to his 1974 work and
each gets its own geological map.

Amaral's vision about the geotectonic processes that shaped the Brazilian part of
the Amazonian Platform suggest a high-grade Archean core, which become remobilised and rejuvenated by later thermal processes. In explaining the origin of the
greenstone belts he follows the gravity-dominated model of Anhaeusser (1975). The
high-grade terrains and the greenstone belts represent the last geosynclinal processes
acting in the Amazonian Platform. Later, periods of stability and reactivations alter-
nate. His models are strongly influenced by the work of Soviet authors working in
the Baltic Shield and elsewhere in Asia. Plate tectonics are still absent in his models.

In 1972 also Colombia started with its own radar and field survey of the natural resources in its Amazonian territory, the Proyecto Radargramétrico del Amazonas (ProRADAM), encompassing the westermost extremities of the Guiana Shield, one of the least known areas of the shield, and partly coinciding with Cordani's Rio Negro province (see Chap. 6). It resulted in a voluminous report *La Amazonia Colombiana y sus recursos*, with 1:200,000 maps of the geology, soils, vegetation and potential land use. The geology was reported by Galvis et al. (1979) and an improved version of that paper was given by Huguett et al. (1979). The main Precambrian units are the Mitú migmatitic complex of high-grade gneisses and younger granites, unfortunately still ascribed by the authors to solid-state potassic metasomatism instead of magmatism. Priem et al. (1982) obtained a Rb–Sr isochron of 1780–1450 Ma for the gneisses, and a conventional U–Pb age of intrusive granites of 1552 Ma, in harmony with earlier ages obtained by Tassinari (1981) from the Brazilian part of his Rio Negro province (see also Kroonenberg 2019).

References

Amaral G (1974) Geologia Pré-Cambriana da região Amazônica. Universidade de São Paulo, Tese de livre-docência, p 212

Amaral G (1984) Províncias Tapajós e Rio Branco. In: de Almeida FFA, Hasui Y (eds) O Précambriano do Brasil, pp 6–35

Anhaeusser CR (1975) Precambrian tectonic environments. Ann Rev Earth Planet Sci 3:31–53

Berrangé J (1977) The Geology of southern Guyana, South America. Institute of Geological Sciences Overseas Memoir 4, pp 112

Galvis J, Huguett A, Ruge P (1979) Geología de la Amazonia Colombiana. Boletín Geológico Ingeominas 22(3):3–86

Gibbs AK (1980) Geology of the Barama-Mazaruni Supergroup of Guyana. Ph.D. thesis, Harvard University, Cambridge, Mass., USA, pp 385

Huguett A, Galvis J, Ruge P (1979) Geología. In: La Amazonia colombiana y sus recursos. Proyecto Radargramétrico del Amazonas, Bogotá, pp 29–92

Issler RS (1975) Geologia do Cráton Guianês e suas possibilidades metalogenéticas. In: Anais Décima Conferência Geológica Interguianas, Belém, Brazil, pp 47–72

Kroonenberg SB (2019) The proterozoic basement of the Western Guiana Shield and the Northern Andes. In: Cediel F, Shaw RP (eds) Geology and tectonics of Northwestern South America. Front Earth Sci, Springer, pp 115–192

Oliveira AS, Fernandes CAC, Issler RS, Abreu AS, de Montalvão RMG, Teixeira W (1975) Geologia. In: Projeto RadamBrasil, vol 9, Folha NA 21 Tumucumaque e parte da Folha NB 21, pp 19–118

Priem HNA, Boelrijk NAIM, Hebeda EH, Verdumen EAT, Verschure RH (1973). Age of the Precambrian Roraima Formation in northeastern South America: Evidence from isotopic dating of Roraima pyroclastic volcanic rocks in Suriname: Geological Society of America Bulletin 84:1677–1684

Priem HNA, Andriessen PAM, Boelrijk NAIM, de Boorder H, Hebeda EH, Huguett A (1982) Geochronology of the Precambrian in the Amazonas region of southwestern Colombia, western Guiana shield. Geol Mijnbouw 61:229–242

Reynolds JH (2002) Origin of the center for Geochronological Research at São Paulo. Revista do Instituto de Geociências—USP, pp 2–1–8

Snelling NJ, McConnell RB (1969) The geochronology of Guyana. Geologie en Mijnbouw 48:201–213

Suszczynski EF (1970) La Géologie et la Tectonique de la Plateforme Amazonienne. Geol Rundsch 59:1232–1253

Tassinari CCG (1981) Evolução geotectônica da Província rio Negro-Juruena na região amazônica. Dissertação de mestrado, Instituto de Geociências, Universidade de São Paulo, p 99

Williams E, Cannon RT, McConnell RB (1967) The folded Precambrian of Northern Guyana related to the geology of the Guiana Shield. Geological Survey of Guyana Records 5, pp 60

Chapter 6
Geochronological Provinces: A Brazilian Point of View

Abstract In Brazil the powerful concept of continental growth of the Amazonian Craton by accretion of discrete geochronological provinces around an Archean core was developed on the basis of a great number of geochronological data, inspired by African and North-American examples. The delimitation of the geochronological provinces and even their overall validity are being discussed even up to the present time.

6.1 The Cordani Model

While Amaral's monumental reconstruction of the lithostratigraphical and geochronological framework of the Amazonian Craton was rather modest as to the geotectonic processes causing its evolution, his erstwhile colleague at USP Umberto Cordani took a different approach with his team. He suggested that tectonic processes in the Proterozoic differed from the plate tectonics processes preponderant in Phanerozoic terrains. Inspired by the work of Alfred Kröner in Africa (Kröner 1977) and using all geochronological data available, now with the internationally accepted Rb-Sr decay constant of $\lambda = 1.42*10^{-11}y^{-1}$, he adopted the notion of continental growth in the sense of Anhaeusser et al. (1969) by the accretion of mobile belts around older nuclei (Cordani et al. 1979; Figs. 6.1, 6.2).

Cordani et al. (1979) subdivide the Amazonian Craton in roughly the same NW-SE belts as Amaral (without citing him here), calling them tectonic provinces, but giving them a completely different meaning. In Cordani's view, the Central Amazonian Province constitutes the oldest cratonic nucleus of the Amazonian Craton around which younger mobile belts were accreted. They publish two Rb-Sr isochrons of the Central Amazonian Province, one of 2700 Ma, based on four data points from the Inajá and Gradaús ranges in the southeastern Guaporé shield, and a younger one of 1800 Ma based on over 40 data points. Nevertheless, they suppose this to be a very old continental nucleus, which saw the presence of later volcanosedimentary sequences of varying ages, including Trans-Amazonian. In spite of the scanty data on which its was based, it remained an influential concept for over 40 years.

S. Kroonenberg, *The Changing Framework of the Guiana Shield*, SpringerBriefs in Earth System Sciences, https://doi.org/10.1007/978-3-031-86334-9_6

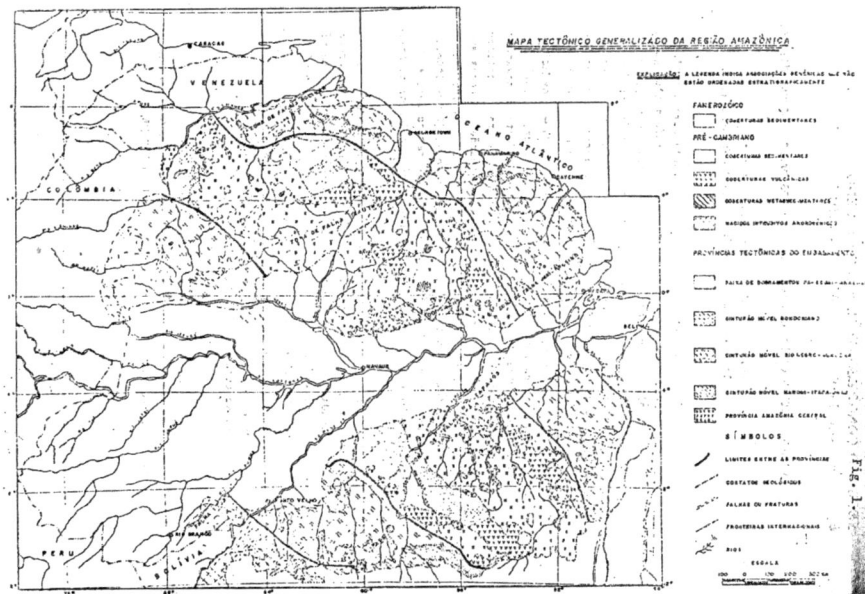

Fig. 6.1 Generalized tectonic map of the Amazonian region: the first 'mobilistic model' of the tectonic provinces of Amazonia, the basis of the continental accretion model of the Amazonian Craton. (Cordani et al. 1979, see also Fig. 6.2)

Along the northern border of the Central Amazonian Province the Maroni-Itacaiúnas mobile belt accreted during the Trans-Amazonian cycle, in their view. This belt encompasses not only the greenstone belts from Venezuela, the Guianas and Amapá, but also the Archean Imataca complex in Venezuela and the Archean Carajás complex in the Guaporé Shield, south of the Amazon River. The Maroni River is the border river between Suriname and French Guiana, the Itacaiúnas River drains the Carajás area. Also here they present two Rb-Sr isochrons, one of 2700 Ma based on over 10 data points, and another one of 1950 Ma based on over 20 data points. They consider the younger felsic volcanic and granitic magmatism in the Central Amazonian Province as a result of a post-tectonic Trans-Amazonian overprint.

Along the southwestern border of the Central Amazonian Province another mobile belt accreted, the Río Negro-Juruena belt, largely in Colombia, southern Venezuela and northwestern Brazil as far as the Guiana Shield is concerned. They present a 1670 Ma Rb-Sr isochron based on about 50 data points for the northern and the southern part of the belt together. Further to the southwest they defines a still younger Rondonian belt with ages around 1100 Ma, of less relevance for the Guiana Shield. Low initial $^{87}Sr/^{86}Sr$ values suggest that accretion of juvenile material to the Amazonian Craton was more important than crustal reworking (see also chapter 10). Which processes could be responsible for the accretion of mobile belts is not discussed.

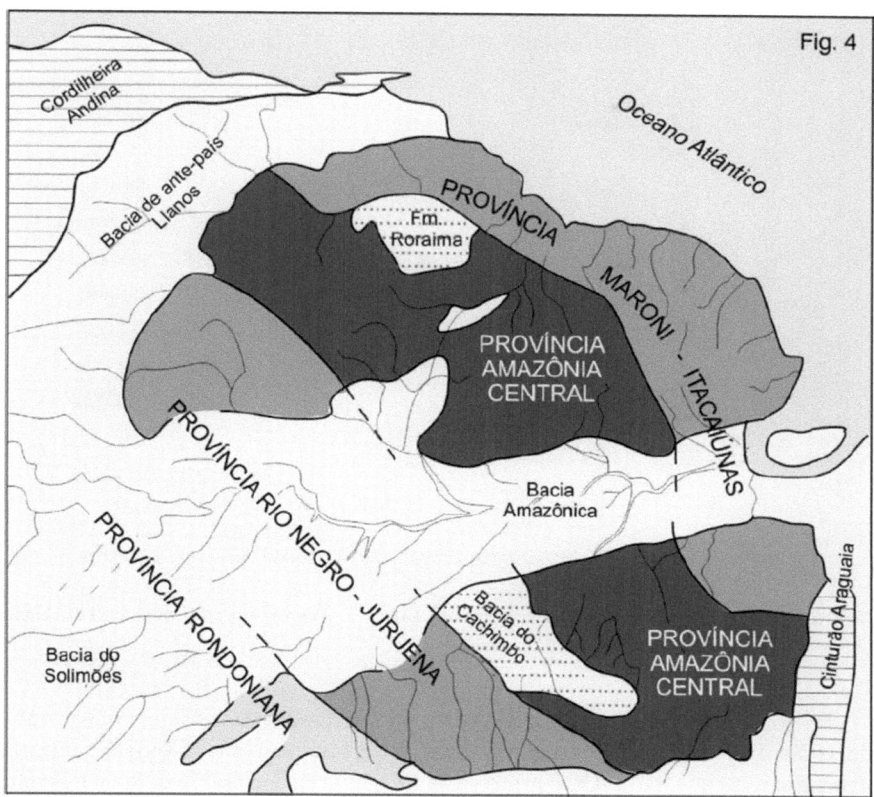

Fig. 6.2 Simplified rendering of Fig. 6.1 by Cordani (2017)

Later Cordani's student and later successor Colombo Tassinari clearly illustrates a role for plate tectonics (Tassinari 1981; Fig. 6.3).

In their paper on the Archean and Early Proterozoic evolution of South America Cordani and Brito Neves (1982) agree that horizontal movements of continental fragments within orogenic belts, and of continental masses in plate tectonic processes, are common at present, and should have been similarly active in the past. The presence of Archean nuclei within the Maroni-Itacaiúnas Province such as Imataca demonstrates the ensialic character of the mobile belt, which was formed, probably in its major part, over pre-existing continental crust (Cordani and Brito Neves 1982) In Cordani et al. (1988) he further elaborates the concept in terms of the growth rates of continents.

It is useful now to follow the fate of Cordani's model of continental accretion around the Central Amazonian Province in Brazil before returning to the development in the other Guianian countries. Anyway, it was ignored in the review paper on the Geology of Brazil by the nestor of Brazilian geology F.F.M. de Almeida et al. (1981), also at USP, in a special issue of *Earth Science Reviews*.

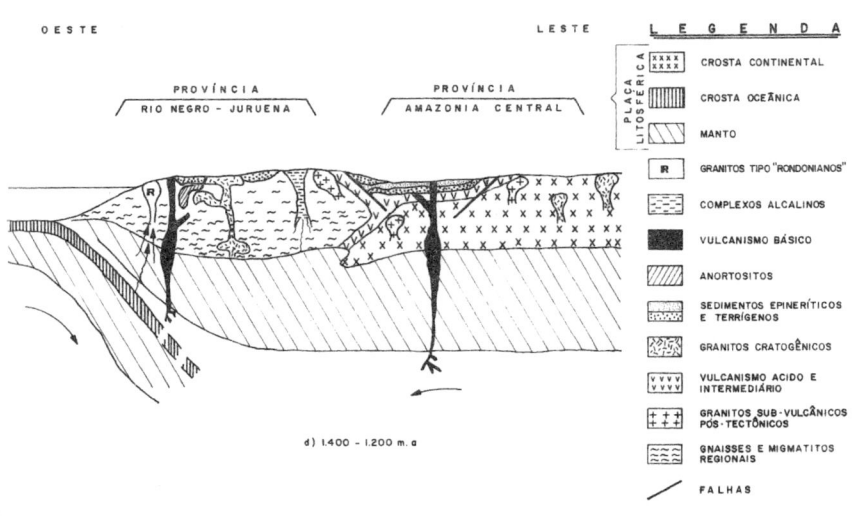

Fig. 6.3 Plate tectonic scheme of the accretion of the Rio Negro-Juruena province to the Central Amazonian Province (Tassinari 1981)

Wilson Teixeira and his colleagues from USP, Tassinari, Cordani, and Kawashita in 1989 reviewed the knowledge obtained in the ten years after Cordani's model in the wake of a successful IGCP project 204 Precambrian Evolution of the Amazonian Region (Teixeira et al. 1989; Fig. 6.4). They state that the present discussion focusses on whether the craton is a large Archean platform, partially reworked and reactivated during the Proterozoic, or whether its evolution is punctuated by episodic crustal accretion during the Proterozoic.

On the basis of an isotopic database of about 3000 age determinations they partition the Amazonian Craton into *geochronological provinces*, a concept already introduced by Cordani et al. (1979). A geochronological province is defined by them as a region, typified by a coherent age pattern over all of its areal extent. They in general maintain the main subdivision of Cordani et al. (1979) in an ancient nucleus, the Central Amazonian Province, surrounded by the accreted early to middle Proterozoic provinces Maroni-Itacaiúnas, Rio Negro-Juruena, Rondonian and Sunsás, though with some modifications: the Archean Carajás area south of the Amazon basin is now transferred from the Maroni-Itacaiúnas Province to the Central Amazonian Province, which may have served to support the thusfar scanty evidence that the latter is an ancient cratonic nucleus. The younger felsic volcanics in the Central Amazonian province are indicated as a platform cover on top of the core of the province. At the other hand, they include the Archean high-grade Imataca belt and the Central Guiana Granulite Belt of Gibbs and Barron (1983) into the Maroni-Itacaiúnas belt adscribed to the Trans-Amazonian Orogeny between 2.1 and 1.9 Ga, rendering it the shape

215

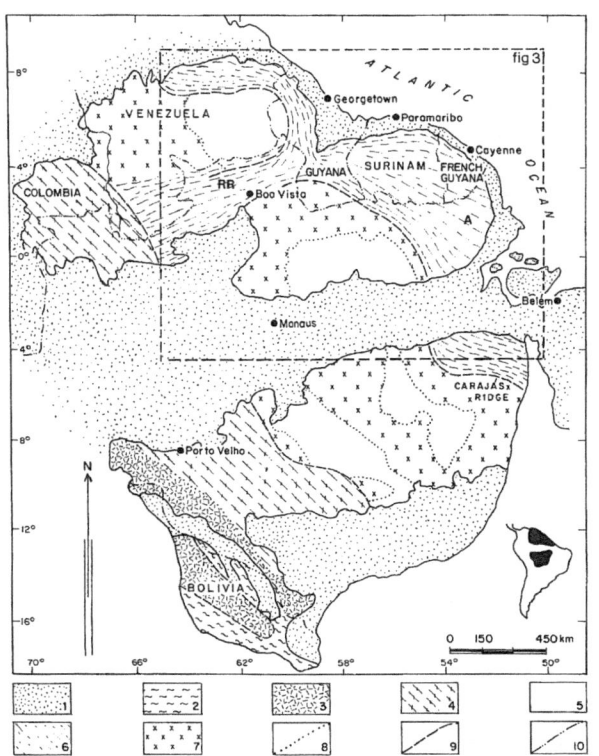

GEOTECTONIC OUTLINE OF THE AMAZONIAN CRATON

Fig. 2. Outline of the geochronological provinces, Amazonian Craton. Key: 1 – Phanerozoic sediments; 2 – Sunsas mobile belt (1.1–0.90 Ga); 3 – Rondonian mobile belt (1.45–1.25 Ga); 4 – Rio Negro–Juruena mobile belt (1.75–1.5 Ga); 5 – Proterozoic platform covers (1.9–1.5 Ga); 6 – Maroni–Itacaíunas mobile belt (2.25–1.9 Ga); 7 – Central Amazonian Province (>2.5 Ga); 8 – transition zone between belts; 9 – approximate contacts between belts; 10 – national boundaries. A - Amapá Territory; RR - Roraima Territory.

Fig. 6.4 The geochronological provinces of the Amazonian Craton (Teixeira et al. 1989). Reproduced with permission from Elsevier

of a triple junction (Fig. 6.4). In the western part of the Guiana Shield they note that the E-W trending Maroni-Itacaiunas belt is truncated by the NW-SE striking Rio Negro-Juruena Province with Rb-Sr ages around 1700 Ma, in accordance with Cordani et al. (1979) and Tassinari (1981).

6.2   The Tassinari and Macambira Model

Another ten years later, in the *Episodes* issue distributed at the International Geological Conference in 2000 in Rio de Janeiro organised by Umberto Cordani and his team (Cordani et al., 2000), Colombo Tassinari of USP and Moacir Macambira of the Universidade Federal do Pará presented a new vision of the accretionary model of the Amazonian Craton originally proposed by Cordani et al. (1979), Tassinari and Macambira (1999). It is mainly based on Rb-Sr, K-Ar and U-Pb zircon ages as well as some Sm-Nd whole-rock ages listed in Tassinari's 'tesis de livre-docência' (Tassinari 1996). The Tassinari and Macambira paper has become one of the most cited articles on the geochronology of the Amazonian Craton, and the accompanying map has been reproduced profusely in the international literature (Fig. 6.5). Their geochronological provinces may include older nuclei, sedimentary basins, younger anorogenic plutons and represent one of more orogenic cycles. The map shows clearly that they consider lithostratigraphy and structure subordinate to the continental growth model. Their concept is further elaborated in Tassinari and Macambira (2004).

The Central Amazonian Province, encompassing a great variety of rocks dated between 3.0–1.6 Ga, is now split into two parts, a part in the northern Guiana Shield, the Roraima Block, and a southern part in the southern Guiana Shield and the Guaporé Shield. However, the database of Tassinari (1996) does not list any Archean ages in the Guiana Shield part of the Central Amazonian Province. That whole area is largely underlain by granitoids and felsic volcanics in different age groups between 2000 and 1600 Ma mainly, according to his Rb-Sr data. The only Archean block in the Central Amazonian Province is the Carajás block in the Guaporé Shield, that originally formed part of the Maroni-Itacaiúnas Province of Cordani et al. (1979). Nevertheless, they consider the whole Central Amazonian Province, north and south, as an old cratonic nucleus, following earlier concepts by Cordani et al. (1979) and Teixeira et al. (1989). They present support for that concept in the Nd isotopic evolution curves of Cordani and Sato (1999) (Fig. 6.6, see also Chapter 10) and the accompanying map (Fig. 6.7).

Cordani and Sato (1999) conclude on the basis of, in their words, 'very scanty data for the Amazonian craton' that In Fig. 6.6. most 'geotectonic units plot in the vicinity of the mantle evolution curve (CHUR), exhibiting positive or slightly negative initial εNd values. This would indicate that juvenile material, formed through mantle differentiation processes, is widespread. This is valid for both the Carajás granitoid rocks of Archean age (CA1 in Fig. 6.6.), as well as for the Early to Mid Proterozoic granitoids belonging to the younger tectonic provinces, where most of the samples yielded Sm-Nd T_{CHUR} model ages practically concordant with the Rb-Sr or U-Pb age values in the same samples' (Cordani and Sato 1999, p 169).

It is apparently on the basis of this scanty evidence, that Tassinari and Macambira (1999) prefer to subdivide the Guiana Shield in 'geochronological provinces' that portray the age of the differentiation of the crustal rocks from the mantle or older basement as in Figs. 6.5 and 6.7, rather than the much younger crystallization age of the rocks at the surface. The victims are of course lithostratigraphy and structure, which would have been the more obvious choices for a geotectonic map. The Brazilian geological survey CPRM did not accept that model in the 1970ies and 1980ies (Santos 2003). In fact, even now there is little support for the hypothesis that the Central Amazonian Province is indeed an old cratonic area around which the younger provinces were accreted. I will come back to that question further below.

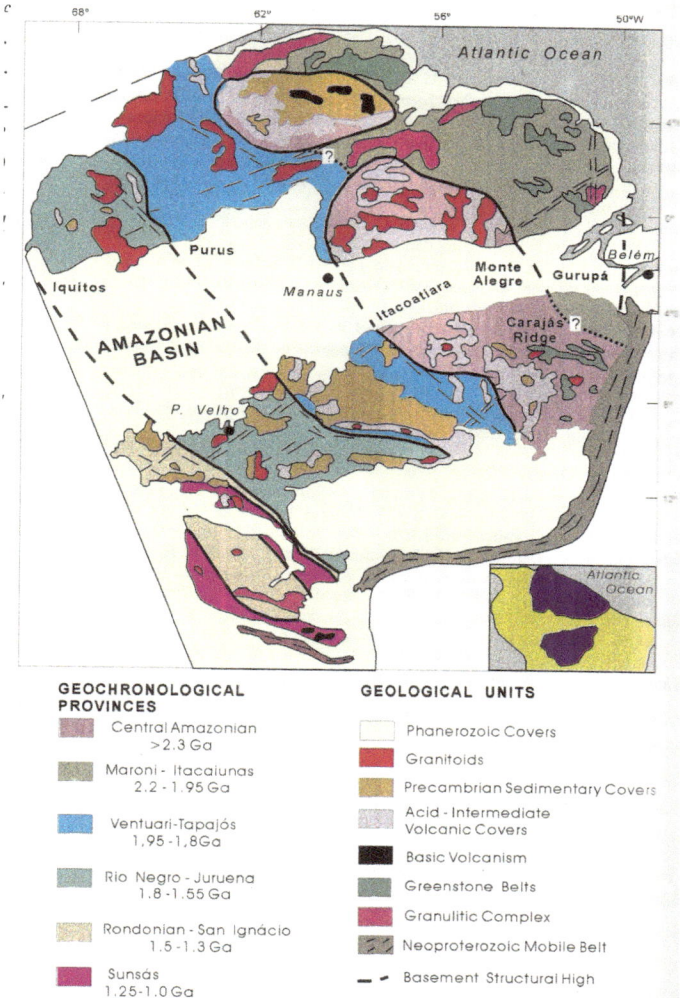

Figure 1 **Sketch map showing the distribution of the Geochronological Provinces and the main lithological associations of the Amazonian Craton, north of South America. (map based on Tassinari, 1996)**

Fig. 6.5 Geochronological provinces of the Amazonian Craton (Tassinari and Macambira 1999). Reproduced with permission Episodes

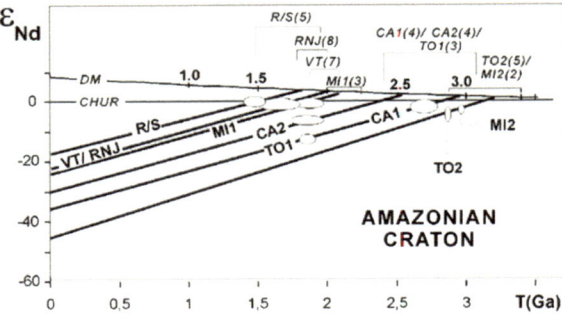

Fig. 6.6 Diagram of Nd isotopic evolution for the Amazonian Craton CA1 and CA2=Central Amazonian province; MI1 and MI2=Maroni-Itacaiunas province; VT=Ventuari-Tapajòs province; RNJ=Rio Negro-Juruena province; R=Rondonian belt; S=Sunsás belt; T01 and T02=Basement inliers within Tocantins belt

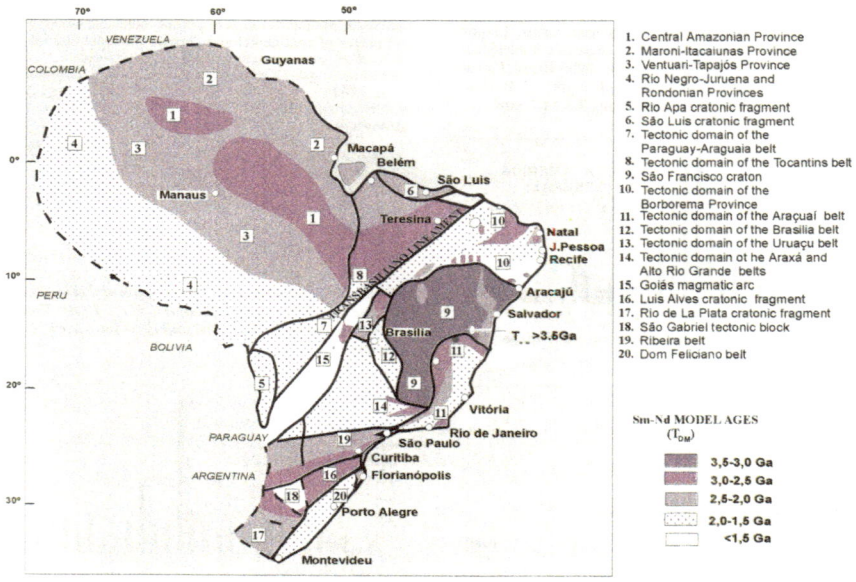

Figure 8 Crustal domains of the South American Platform.

Fig. 6.7 Crustal domains of the South-American platform (Cordani and Sato 1999). Reproduced with permission from Episodes

The Maroni-Itacaiúnas province retained its name in Tassinari and Macambira (1999) even though the Itacaiúnas River drains the Carajás area which is now incorporated in the Central Amazonian Province. It encompasses both the 2.2–1.9 Ga greenstone belts of the northern Guiana Shield as well as the high-grade belts, of which Imataca in Venezuela (Tassinari et al. 2004) and Cupixí in Amapá present

Archean ages. They introduce a new geochronological province between the Central Amazonian Province and the Rio Negro-Juruena Province, mainly consisting of calcalkaline granitoid rocks, called Ventuari-Tapajós, 1.95–1.8 Ga, which equally crosses the Amazon basin. Further west the Rio Negro-Juruena borders the Ventuari-Tapajós Province (Tassinari et al., 1996, 2000). The still younger Rondonian-San Ignacio and Sunsás provinces further west do not crop out in the Guiana Shield.

Tassinari and Macambira (1999) suggest that part of the Maroni-Itacaiúnas Province and the whole of the Ventuari-Tapajós province, as well as the younger provinces in the Guaporé Shield, evolved from the addition of juvenile, mantle-derived magmas, while part of the Maroni-Itacaiunas Province might be associated with the recycling of older continental crust. They estimate that 30% of the Amazonian Craton was derived from the mantle in the Archean and 70% in the Proterozoic. The younging of ages from NE-SW would support their hypothesis of lateral crustal growth during the Paleo-and Mesoproterozoic for the Amazonian Craton.

In their comprehensive paper on the evolution of the Amazonian Craton Cordani and Teixeira (2007) retain the same geochronological provinces of Tassinari and Macambira (1999), even though they admit that Carajás is the only Archean area in the Central Amazonian Province. Subduction-driven 'soft-collision/accretion' would be responsible for continental growth around the Central Amazonian Province.

6.3   The Santos Model

However, since then there has been an enormous upsurge of new, much more precise geochronological analytical methods, notably the U-Pb systematics of individual zircon crystals, such as LA-ICPMS and U-Pb SHRIMP (sensitive, high-resolution ion-microprobe). Zircons have a higher blocking temperature than the K-and Rb-bearing minerals such as biotite, muscovite and hornblende, and in magmatic rocks usually reflect the age of crystallisation. Therefore the resulting zircon ages are generally older than the Rb-Sr isochrons, whole rock data and individual mineral ages. Moreover, the error margins are much smaller than in the Rb-Sr data. That has in general led to discarding the old age data, redating the samples on the basis of their zircons, and at present only zircon U-Pb data are regarded as publishable.

In Brazil, João Orestes Schneider Santos of the Brazilian GeologicaL Survey CPRM was the first to apply single-crystal zircon U-Pb data to some of the accretionary belts of Tassinari and Macambira, using several local, US and Australian laboratories (Santos et al. 2000). This led to another subdivision of the tectonic-geochronological provinces of the Amazonian Craton (Fig. 6.8). He separated the Archean Imataca, Amapá and Carajás domains, renamed the Ventuari-Tapajós domain in Tapajós-Parima and separated the Rio Negro in the Guiana Shield from the Juruena domain in the Guaporé shield and added Rondoniana to the latter. He also introduces a NE-SW running K'Mudku Belt across the Guiana Shield which will be discussed below. The range of ages in those domains are now base on conventional U-Pb, SHRIMP data and Sm-Nd model ages (Fig. 6.9), but the principle of continental accretion is not challenged. He suggests that the Uatumã young felsic volcanics and shallow granitoid intrusions within the Central Amazonian Province are juvenile anorogenic magmatism triggered by low-angle slab subduction beneath a westward continuation of the Carajás Archean block. Santos et al. (2003) also obtained the age of the Roraima Formation using U-Pb SHRIMP dating of zircon from intercalated tuff layers at 1873 ± 3 Ma.

In his 2003 paper (Santos 2003) he gave a clear picture of how the subdivision of the Brazilian part of the Amazonian Craton has changed in 25 years (Fig. 6.10).

His final scheme is depicted in Santos et al. (2006a, Fig. 6.11). The K'Mudku collisional belt already introduced in the map of Santos et al. (2000) is supposed to be the a collisional feature between the western and eastern part of the Guiana

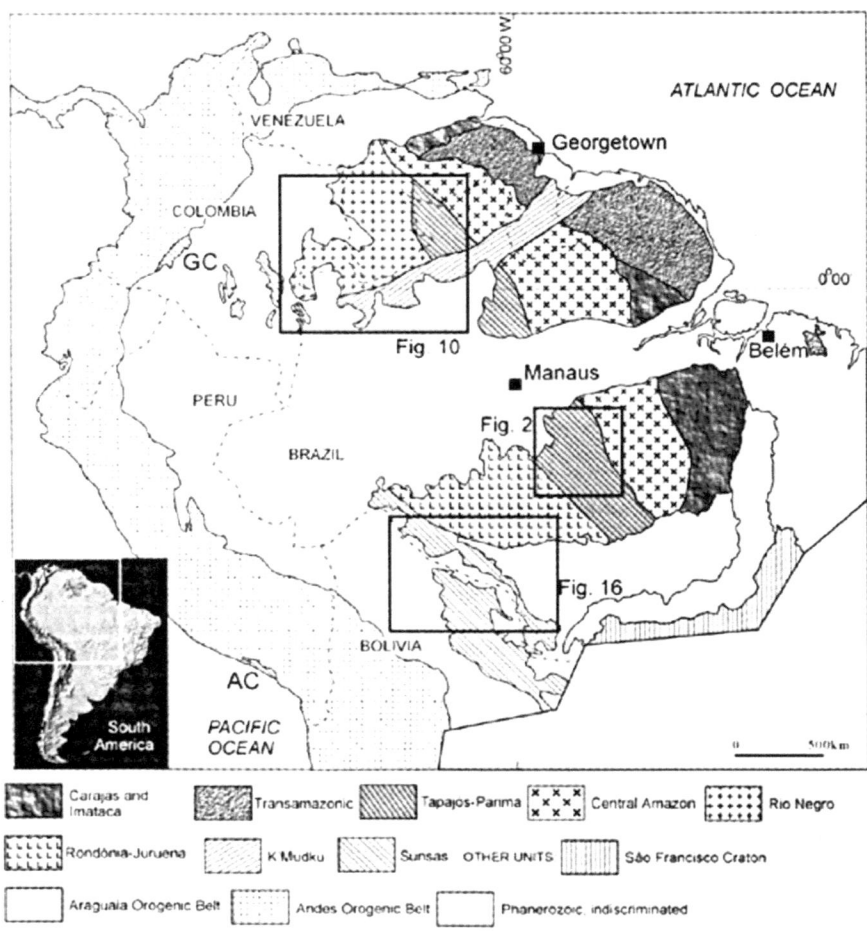

Fig. 6.8 The subdivision in geochronological provinces according to Santos et al. (2000)

Table 1.   Major tectonic, structural and isotopic characteristics of the Provinces of the Amazon Craton.

Province	Main Tectonic Trend	Dominant Process	ε_{Nd}[#]	T_{DM} (Ga)[#]	U-Pb ages[#]
Sunsas	N 40° W	collisional	-4.46 · +6.25	1.93-1.52	1.33 - 0.99
Rondônia-Juruena	N 70° W / E-W	juvenile	· 1.65 / + 3.81	2.18 -1.68	1.76 - 1.47
Rio Negro	N-S				
	N 40° W	collisional	- 5.00 / + 2.98	2.42-1.88	1.86 - 1.52
Central Amazon	NNW	underplating	- 14.39 / +1.52	2.85-2.41	1.88 - 1.70
Tapajós-Parima	N 30° W	juvenile	- 2.38 / + 3.51	2.26-2.06	2.10 - 1.87
				2.32-2.07	
Transamazonic	N 50°-70° W	juvenile	+ 0.20 / + 3.83	3.10*-3.06*	2.25 - 2.00
Carajás	N 70° W	juvenile	- 1.25 / + 5.30	3.10 – 2.51	3.10 - 2.53

- Detail on Table 3 ;   n.a.= not available.;   (*) Cupixi Domain.

Fig. 6.9   Structure and ages of the geochronological provinces in Brazil distinguished by Santos et al. (2000)

Shield (Santos et al. 2006b). It is useful to describe the history of this feature in some detail, because Santos' concept has led to a proliferation of papers which follow this questionable concept, e.g. Souza et al. (2015).

K'Mudku is the name given by Irish-Guyanese geologist Chris Barron to a prominent mylonite zone in southern Guyana in his presentation to the 7th Guiana Geological conference in Paramaribo in November 1966. It was published in an internal Guyanese report (Barron 1969), of which only a very short abstract appeared in the proceedings of the conference as Barron (1969). The mylonite zone had already been remarked by earlier geologists, and McConnell et al. (1964) determined its age between 1300 and 1100 Ma on the basis of K-Ar and Rb-Sr mineral data, also mentioned at the same 1966 conference by Snelling and McConnell (1969). Sobharam Singh, then director of the Guyana Geological Survey, showed the mylonite zone on the map accompanying his PhD thesis (Singh 1966; Fig. 6.12) including the location of the K'Mudku mountain where it is well exposed (Berrangé 1977). Singh remarked the absence of offset along the mylonite zone, as is also evident from his map.

In the meantime, also Harry Priem et al. (1966) had noticed young Rb-Sr and K-Ar biotite ages around 1300 Ma from a granodiorite in southwestern Suriname, but initially he tentatively ascribed it to heating by a nearby dike. Priem et al. (1967), citing Snelling and McConnell (1969) recognised it as a metamorphic event, and introduced the name Nickerie Metamorphic Episode for Suriname in Priem et al. (1968a), after the Nickerie district in western Suriname to where the effect appeared to be limited. He described it more extensively in Priem et al. (1971). So the name K'Mudku should have priority for this event (Berrangé 1977). As cited above, Martín-Bellizzia (1972) referred to it as Orinoquense event.

Santos et al. (2006b) starts by erroneously alleging that Barron (1969) considered the K'mudku mylonite as evidence of a collision. His view is coloured by the concepts of Anhaeusser et al. (1969), that granulites represent younger mobile belts along older cratons. His 2006b map includes the granulites of the Kanuku Mountains in Guyana and Bakhuis Mountains in Suriname, for which Trans-Amazonian metamorphic ages around 2.07–2.03 Ga had already been published long ago (e.g. De Roever et al. 2003, see also Nanne et al. 2020) and from which Nickerie (K'Mudku)

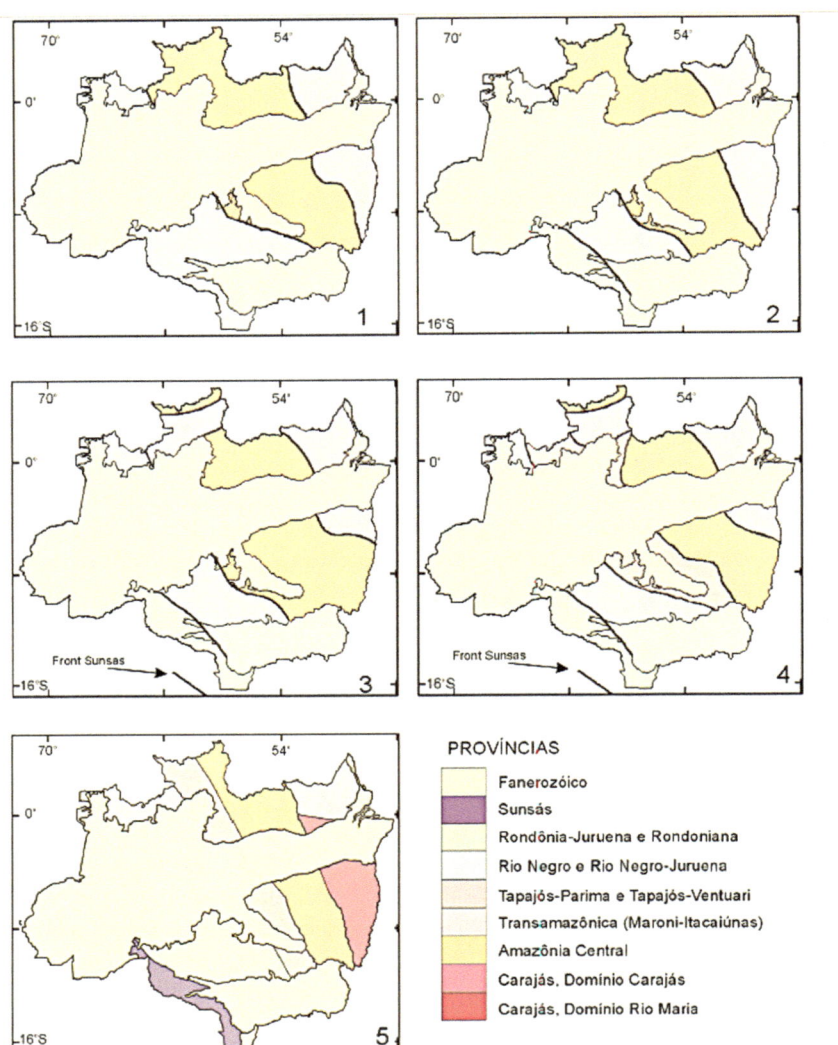

Fig. 6.10 Development of geochronological provinces in Brazil (Santos 2003). Note that diagram 5 (Santos et al. 2000) in this figure differs from the map in Santos et al. (2000) cited above (Fig. 6.8)

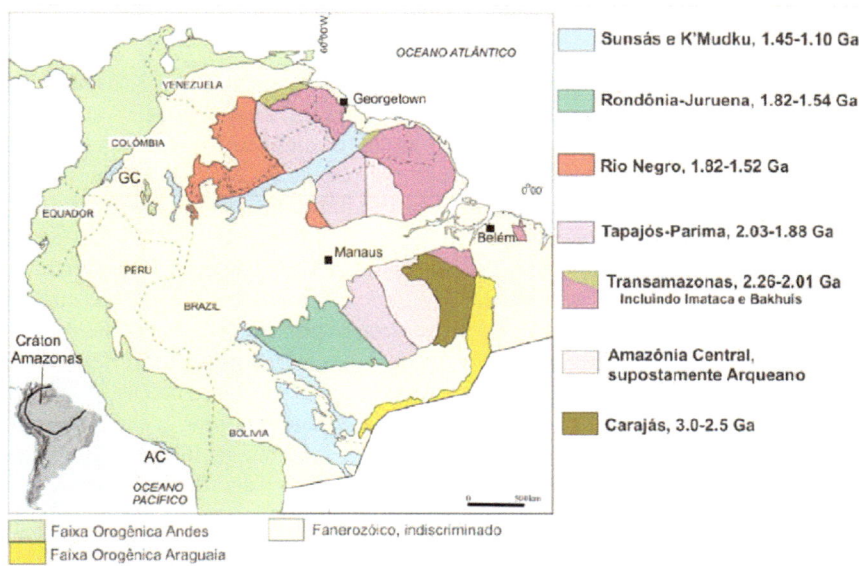

Fig. 6.11 Santos et al. (2006a) final subdivision of the Amazonian Craton

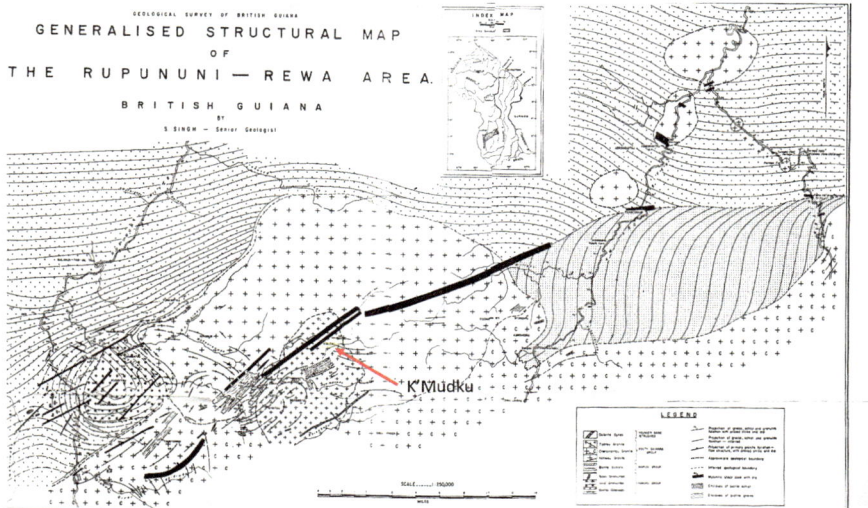

Fig. 6.12 The K'Mudku mylonite zone in Guyana, the location of the K'Mudku Mountain (red arrow), and the lack of set-off along the shear zone (modified after Singh 1966)

ages only represent a late thermal recrystallisation, not higher than in the greenschist or prehnite-pumpellyite facies (Bosma et al. 1984).

Santos et al. (2006b) cites several Ar-Ar ages from micas in his K'Mudku zone in Brazil between 1185 and 1337 Ma in rocks with Trans-Amazonian crystallization ages. He also presents allegedly high-grade metamorphic titanite ages between 1490 and 1147 Ma. However, titanite can only be considered as an amphibolite-facies metamorphic mineral if it is extremely aluminous (Aleinikoff et al. 2022). Without mineral analysis his interpretation that titanite represents high-grade metamorphism is unsubstantiated, and contradicts the low-grade metamorphic characteristics recorded elsewhere in the shield. It is important to note that low-grade thermal mica age resetting occurs everywhere in the western part of the Guiana Shield, also in undeformed rocks far removed from mylonite zones, e.g. Pinson et al. (1962), Priem et al. (1971, 1977, 1982). Santos (2006b) considers the K'Mudku belt as an intracontinental reflex of the Sunsás Orogeny at the southwestern border of the Amazonian Craton. What this 'intracontinental reflex' means remains unclear; the 2006a map suggests an intracontinental collision between the western and the eastern part of the shield, which is certainly not the case. Most authors at present consider the mylonitization and the low-grade thermal age resetting in the Guiana Shield as an effect of the Grenvillian collision of Laurentia and Amazonia (Kroonenberg 1982; Cordani et al. 2010) termed Putumayo Orogeny by Ibáñez-Mejía et al. (2011).

References

Aleinikoff JN, Wintsch JP, Fanning CM, Dorais MJ (2002) U-Pb geochronology of zircon and polygenetic titanite from the Glastonbury complex, connecticut, USA: an integrated SEM, EMPA, TIMS, and SHRIMP study. Chem Geol 188:125–147

Anhaeusser CR, Mason R, Viljoen M, Viljoen RP (1969) A reappraisal of some aspects of Precambrian Shield geology. Geol. Soc. Amer. Bull 80:2175–2200

de Almeida FFM, Hasui Y, De Brito Neves BB, Fuck RA (1981) Brazilian structural provinces: an introducion. Earth-Sci Rev 17:1–29

Barron CN (1969) Notes on the stratigraphy of Guyana. Proceedings Seventh Guiana Geological Conference, Paramaribo, 1966. Records Geological Survey Guyana 6, II: 1–28

Berrangé J (1977) The geology of southern Guyana, South America. Inst Geol Sci Overseas Mem 4:112

Bosma W, Kroonenberg SB, van Lissa R, Maas K, de Roever EWF (1984) Explanation to the geological map of suriname 1:500,000. Mededelingen Geologisch Mijnbouwkundige Dienst Van Suriname 27:31–82

Cordani UG, Tassinari CCG, Teixeira W, Basei MAS, Kawashita K (1979) Evolução tectônica da Amazonia com base nos danos geocronológicos. 2ndo Congreso Geológico Chileno 139–148

Cordani UG, de Brito Neves BB (1982) The geologic evolution of South America during the Archaean and Early Proterozoic. Revista Brasileira De Geociências 12(1–3):78–88

Cordani UG, Teixeira W, Tassinari CCG, Kawashita K, Sato K (1988) The growth of the Brazilian Shield. Episodes 11(3):163–167

Cordani UG, Sato K (1999) Crustal evolution of the South American Platform, based on Nd isotopic systematics on granitoid rocks. Episodes 1999:167–173

Cordani UG (2017) História geológica do Craton Amazônico Anais do 15 Simpósio de geologia da Amazonia, Belém 11–16

Cordani UG, Sato K, Teixeira W, Tassinari CCG, Basei MAS (2000) Crustal evolution of the South American Platform. In: Cordani U.G. et al. (Eds.) Tectonic evolution of South America, p. 19–40

Cordani UG, Teixeira W (2007) Proterozoic accretionary belts in the Amazonian Craton.in: R.D. Hatcher, UG et al (2007) 4D framework of continental crust. Geol Soc Amer Memoir 200:297–320

Cordani UG, Fraga LM, Reis N, Tassinari CCG, Brito-Neves BB (2010) On the origin and tectonic significance of the intra-plate events of Grenvillian-type age in South America: A discussion. J South Am Earth Sci 29:143–159

De Roever EWF, Lafon JM, Delor C, Rossi P, Cocherie A,.Guerrot C, Potrel A (2003) The Bakhuis Ultra-high temperature granulite belt : I Petrological and geochronological evidence for a counterclockwise P-T path at 2.07–2.05 Ga. Géologie de la France 2003, 2–3–4:175–205

Gibbs AK, Barron CN (1983) The Guiana Shield reviewed. Episodes 1983:7–14

Ibáñez-Mejía M, Ruiz J, Valencia VA, Cardona A, Gehrels GE, Mora AR (2011) The Putumayo Orogen of Amazonia and its implications for Rodinia reconstructions: New U-Pb geochronological insights into the Proterozoic tectonic evolution of Northwestern South America. Precambr Res 191:58–77

Kröner A (1977) The Precambrian geotectonic evolution of Africa: plate accretion versus plate destruction.Precambrian research 4:163–213

Kroonenberg SB (1982) A Grenvillian granulite belt in the Colombian Andes and its relation to the Guiana Shield. Geologie and Mijnbouw 61:325–333

Martín-Bellizzia C (1972) Paleotectónica del Escudo de Guayana. Memoria de la 9na conferencia geológica Inter- Guayanas. Boletín de Geología (Caracas), Publicación especial No. 6:251-304

McConnell RB, Cannon RT, Williams E, Snelling NJ (1964) A new interpretation of the geology of British Guiana. Nature 204:115-118

Nanne JAM, de Roever EWF, de Groot K, Davies GR, Brouwer FM (2020) Regional UHT metamorphism with widespread, primary CO2-rich cordierite in the Bakhuis Granulite Belt, Surinam: A feldspar thermometry study. Precambr Res 350:105894

Pinson WH, Hurley PM, Mencher E, Fairbairn HW (1962) K-Ar and Rb-Sr ages of biotites from Colombia, South America. Geol Soc Am Bull 73:807–910

Priem HNA, Andriessen PAM, Boelrijk NAIM, de Boorder H, Hebeda EH, Huguett A (1982) Geochronology of the Precambrian in the Amazonas region of southwestern Colombia, western Guiana shield. Geol Mijnbouw 61:229–242

Priem HNA, Boelrijk NAIM, Verschure RH, Hebeda EH (1966) Isotopic age determinations on Surinam rocks. Geol Mijnbouw 45:16–19

Priem HNA, Boelrijk NAIM, Verschure RH, Hebeda EH (1967) Isotope age determinations on Surinam rocks, 2. Geol Mijnbouw 46:482–486

Priem HNA, Hebeda EH, Boelrijk NAIM, Verschure RH (1968) Isotope age determinations on Surinam rocks, 3. Proterozoic andc Permo-Triassic basalt magmatism in the Guiana Shield. Geol Mijnbouw 47:17–20

Priem HNA, Boelrijk NAIM, Hebeda EH, Verdurmen EAT, Verschure RH (1971) Isotopic ages of the Trans-Amazonian acidic magmatism and the Nickerie Episode in the Precambrian basement of Surinam, South America. Geol Soc Am Bull 82:1667–1680

Priem HNA, Boelrijk NAIM, Hebeda EH, Kroonenberg SB, Verdurmen EAT, Verschure RH (1977) Isotopic ages in the high grade metamorphic Coeroeni Group, southwestern Suriname. Geol Mijnbouw 56(155):160

Santos JOS (2003) Geotectônica dos Escudos das Guianas e Brasil-Central. Geotectonics of the Guyana and Central Brazilian Shields. In: Bizzi L.A, Schobbenhaus C, Vidotti R.M, Gonçalves E.J.H (eds.) Geologia, Tectônica e Recursos Minerais do Brasil 169–226

Santos JOS, Hartmann LA, Gaudette HE, Groves DI, McNaughton NJ, Fletcher IR (2000) A new understanding of the provinces of the Amazon Craton based on integration of field mapping and U-Pb and Sm-Nd geochronology. Gondwana Res 3:453–488

Santos JOS, Potter PE, Reis NJ, Hartmann LA, Fletcher IR, McNaughton NJ (2003) Age, source and regional stratigraphy of the Roraima Supergroup and Roraima-like outliers in northern South America based on U-Pb geochronology. Geol Soc Am Bull 115:331–348

Santos JOS, Hartmann LA, Faria MS, Riker SR, Souza MM, Almeida ME, McNaughton NJ, (2006a) Compartimentação do Cráton Amazonas em províncias: avanços ocorridos no período 2000–2006. Simpósio de Geologia da Amazônia, vol. 9, Sociedade Brasileira de Geologia, Belém, Brazil, Resumos Expandidos, CD ROM

Santos JOS, Faria MS, Riker SR , Souza MM, Hartmann LA, Almeida ME , McNaughton NJ, Fletcher IR (2006b) A faixa colisional K'mudku (idade Grenvilliana) no norte do Cráton Amazonas: reflexo intracontinental do Orógeno Sunsás na margem ocidental do cráton. Simpósio de Geologia da Amazônia, vol. 9, Sociedade Brasileira de Geologia, Belém, Brazil, Resumos Expandidos, CD ROM

Singh S (1966) Geology and petrology of part of the Guiana Shield in the South Savanna-Rewa area of Guyana. Geol Surv Dep, Georg, Guyana, Bull 37:127

Snelling NJ, McConnell RB (1969) The geochronology of Guyana. Geologie en Mijnbouw 48:201–213

Souza VS, de Souza AGH, Dantas EL, Valério CS (2015) K´Mudku A-type magmatism in the southernmost Guyana Shield, central-north Amazon Craton (Brazil): the case of Pedra do Gavião syenogranite. Braz J Geol 45:293–306. https://doi.org/10.1590/23174889201500020008

Tassinari CCG (1981) Evolução geotectônica da Província rio Negro-Juruena na região amazônica. Instituto de Geociências, Universidade de São Paulo, Dissertação de mestrado, p 99

Tassinari CCG (1996) O mapa geocronológico do Cráton Amazônico no Brasil: Revisão dos dados isotópicos. Tesis de livre-docencia Universidade de São Paulo 257

Tassinari CCG, Cordani UG, Nutman AP, van Schmus WR, Bettencourt JS, Taylor PN (1996) Geochronological systematics on basement rocks from the Rio Negro-Juruena Province Amazonian Craton, and tectonic implications. Int Geol Rev 38:161–175

Tassinari CCG, Macambira MJB (1999) Geochronological provinces of the Amazonian Craton. Episodes 22(3):174–182

Tassinari CCG, Bettencourt JS, Geraldes MC, Macambira MJB, Lafon J-M (2000) The Amazonian Craton. In: Cordani, U.G et al. (eds.) Tectonic evolution of South America. 31st International Geological Congress, Rio de Janeiro, Brazil: 41–95

Tassinari CCG, Macambira MJB (2004) A evolução tectônica do Cráton Amazônico. In: Geologia do continente sul-americano: evolução da obra de Fernando Flávio Marques de Almeida. São Paulo: Beca 471–485

Tassinari CCG, Munhá JMU, Teixeira W, Palacios T, Nutman A, Sosa SC, Santos AP, Calado BO (2004) The Imataca Complex, NW Amazonian Craton, Venezuela: crustal evolution and integration of geochronological and petrological cooling histories. Episodes 27:3–12

Teixeira W, Tassinari CCG, Cordani UG, Kawashita K (1989) A review of the geochronology of the Amazonian Craton: Tectonic Implications. Precambr Res 42:213–227

Chapter 7
Greenstone Belts in the Guianas

Abstract In spite of the enormous amounts of data and their geotectonic interpretation of the Amazonian Craton produced by the Brazilians, geological research in Guyana, Suriname, French Guiana, Venezuela and the Brazilian state of Amapá went largely its own way, focusing rather on the stratigraphy, structure and mineral deposits in the greenstone belts along the Atlantic coast than on their role in the continental accretion scenario. The Trans-Amazonian Orogeny between 2.26-1.98 Ga was the main tectonic event shaping their part of the shield.

When the Brazilians presented their amazing RadamBrasil data and new concepts about the subdivision of the craton in geochronological provinces, the effect of these ideas in the Guianas and Venezuela was limited, partly because almost all of their territories fell indiscriminatorily within the Maroni-Itacaiúnas province, covered by greenstone belts. Moreover, after the successful Tenth Inter Guiana Geological Conference in 1975 in Belém the conference series was interrupted for more than 40 years. In the meantime each Guianian country separately developed its own ideas about the structure of the Guiana Shield.

7.1 Guyana

In Guyana the landmark thesis of Allan Gibbs about the Barama-Mazaruni greenstone belt of northern Guyana appeared (Gibbs 1980), and in 1983 he published a review of the geology of the Guiana Shield in *Episodes* (Figs. 7.1, 7.2), together with his father-in-law Chris Barron, an Guyanese-Irish geologist who had worked in Guyana since 1954 (Gibbs and Barron 1983). They present a well-documented scheme of major Precambrian Units and events, but although they refer to papers by Cordani and Santos, they do not mention the geochronological provinces nor comment upon their accretionary hypothesis. They show Middle Proterozoic acid volcanics, labeled Uatumã Supergroup, and granitoids as the dominant lithology in the area of Tassinari's Central Amazonian Province. They consider the greenstone

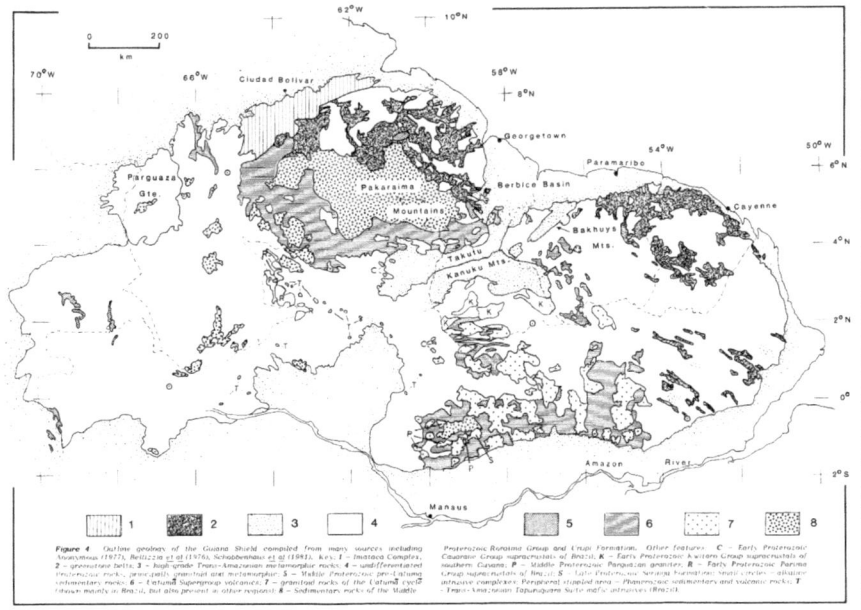

Fig. 7.1 Outline geology of the Guiana Shield, (Gibbs and Barron 1983) Reproduced with permission episodes

belt as of ensimatic origin. Both the Suriname and Guyanese reviews were essentially based on Rb–Sr isotope data. Their 1983 paper was the runner-up to their well-known 1993 Oxford book *Geology of the Guiana Shield* (Gibbs and Barron 1993). Gibbs' brilliant career unfortunately came to a dramatic end by a fatal traffic accident in 1988.

Nevertheless Chris Barron succeeded in finishing the book after Gibbs' death, including a coloured map showing the major lithological units and structural characteristics, with much more detail than the 1983 map, especially in the western part of the Guiana Shield. There are also some striking differences with the 1983 map, e.g. in portraying the high-grade belts in the central parts of the shield. The book contains a lot of useful information and a very extensive bibliography, but the last references are around 1985 (sometimes still labeled 'in press'), and some chapters fall short of the high quality of others. It is useful to read the review of the book by N.J. Snelling (1995).

In the book the authors subdivide the Guiana Shield in geological provinces of their own (Fig. 7.3), which do not coincide with the Brazilian geochronological provinces nor refer to them, but do not play an important role in the book either. Their stratigraphy, treated in great detail across the different countries, largely follows that of the 1983 paper (Fig. 7.2). They consider the whole of the shield as of juvenile origin except for the Imataca Archean core. The three more or less parallel greenstone belts in northwestern Guyana, from north to south the Barama, Cuyuni

Fig. 7.2 Stratigraphic table of the Guiana Shield (Gibbs and Barron 1983). Reproduced with permission episodes

Table I. Major Precambrian units and events of the Guiana Shield.

Key to localities: V - Venezuela, G - Guyana, S - Suriname, F - French Guiana, B - Brazil, C - Colombia; c - central, n -northern, e -eastern, s - southern, w - western portions of the shield within the various countries. Boundary between Early and Middle Proterozoic taken as 1.9 Ga.

LATE PROTEROZOIC

Tectonothermal events (1.3-0.9 Ga)	Nikerie metamorphism, K'Mudku mylonites, Jari-Falsino and Orinoquan events
Alkaline complexes: (ages uncertain)	Cerro Impacto (cV), Muri Mountains (sG, nB), Seis Lagos? (wB), Maracanai? (sB)
Volcanics:	Seringa Fm. (sB), Piraparana Fm.(C)

MIDDLE PROTEROZOIC

Parguazan granites: (1.55-1.45 Ga)	Parguaza (wV), Surucucus (nB), Abonari (sB), Tiquié (wB)
Mafic intrusives: (1.7-1.5 Ga)	Avanavero Suite (V, G, S, B), Quarenta Ilhas (sB), Pedra Preta (nB)
Sedimentary rocks: (1.65-1.75 Ga)	Roraima Gp. (V, G, S, B), Urupi Fm. (sB)
Felsic volcanics (1.75-1.90 Ga)	Uatumã Supergroup (V, G, S, B): Cuchivero (V), Burro-Burro and Kuyuwini Gps. (G), Dalbana Fm. (S), Surumu Fm. (nB), Iricoume Fm. (sB)
and associated granites:	Cuchivero (nV), Saracura (nB), Mapuera (sB)
Pre-Uatumã sedimentary rocks:	Muruwa Fm. (cG), Ston Fm. (wS), Tunui Gp.? (wB), Cinaruco Fm.? (wV), Los Caribes Fm.? (eV), La Pedrera Fm.? (C), La Quina Fm.? (wV)

EARLY PROTEROZOIC

Trans-Amazonian granites: (2.05-2.15 Ga)	South Savanna (sG), Kartabu (cG), Gran Rio (S), Agua Branca (sB)
Trans-Amazonian mafic intrusives:	Tapuruquara Suite (cB) Appinitic Suite (sG), De Goeje Suite (S), San Juan de Manapiare (sV)
Trans-Amazonian metamorphic complexes:	Ile de Cayenne (F), Supamo (cV), Bartica (cG), Kanuku (sG), Guianense Complex (B), Mitu Complex? (C).
Greenstone belts: (2.25 Ga)	Pastora Sgp. (V), Barama-Mazaruni Sgp. (G), Marowijne Gp. (S), Paramacca Sgp. (F), Vila Nova (eB)
Other meta-supracrustal units	Coeroeni and Fallawatra Gps. (S), Cauarane (cB), Kwitaro and Kanuku Gps. (sG), Parima Gp. (nB)

ARCHAEAN

La Ceiba migmatite (nV) 2.7 Ga
Gurian Orogeny (pre 3.0 Ga)
Imataca Complex (nV) (pre-3.5 Ga protolith)

and Mazaruni belts in Guyana show similar stratigraphy, starting with basaltic and komatiitic volcanics, followed by andesitic to rhyolitic volcanics, volcaniclastics and chemical sediments, with turbiditic greywackes, phyllites and conglomerates in the upper parts. No unconformities have been found within them. There were thought to have formed during an Early Proterozoic Arc stage. Gibbs (1980) and Gibbs and

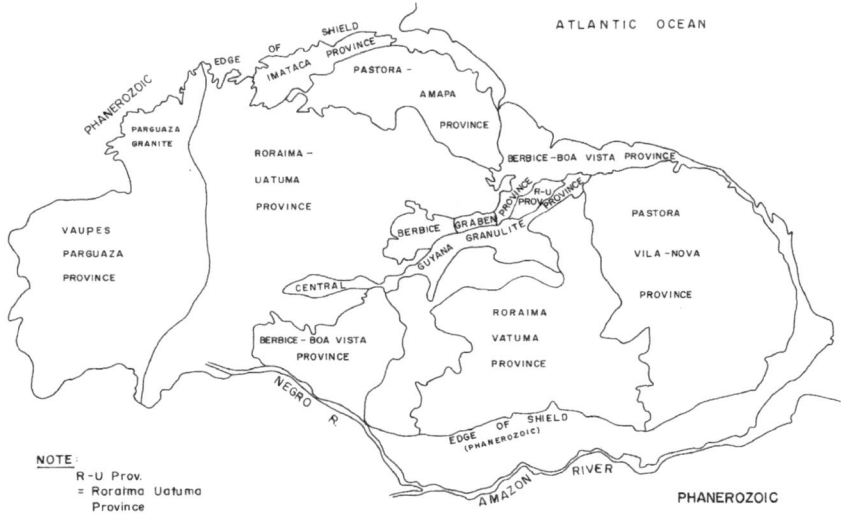

Fig. 7.3 Geological provinces in the Guiana Shield (Gibbs and Barron 1993)

Olszewski (1982) obtained a 2245 Ma conventional U-Pb age on detrital zircons from greywackes, and considered them to date the volcanics from which they were supposed to have been derived.

Folding, metamorphism and plutonism in the whole northern Guiana Shield took place around 2.1–2.0 Ga in the Trans-Amazonian Tectonothermal Episode. Anorogenic granites and felsic volcanics in the shield represent a platform stage between 1.85 and 1.65 Ga, all ages still based on Rb-Sr systematics (Fig. 7.4). They support Tassinari's (1981) hypothesis that the younger westernmost [Rio Negro] province, here called Vaupés-Parguaza province, originated by subduction under a Trans-Amazonian plate in the east.

Tedeschi et al. (2020) describe, after an excellent review of the geology of the Trans-Amazonian greenstone belt, a succession of (partly magnesian) tholeiitic basalts, immature sandstones and conglomerates belonging to the Barama-Mazaruni Supergroup, intruded by quartz monzonites, granodiorites, andesitic and rhyolitic dykes in the Karouni gold deposit in Guyana. Apart from an andesite dyke at 2.15 Ga most intrusives show ages between 2.09 and 2.12 Ga, in the same interval as the Omai intrusives in Guyana (Norcross et al. 2000). Typical older TTG plutons as recorded in French Guiana and Suriname have not been observed in the Karouni area nor elsewhere in Guyana (Tedeschi et al. 2020).

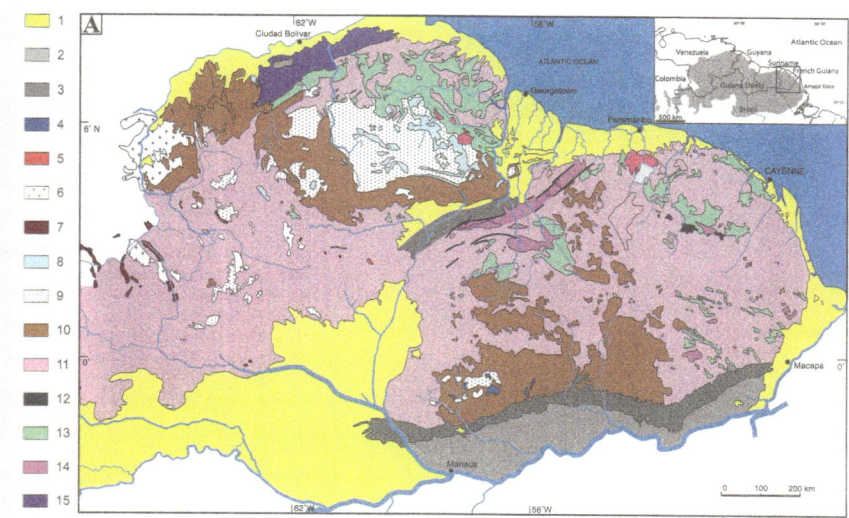

Fig. 7.4 Geological map of the Guiana Shield (Gibbs and Barron 1993, redrawn by Delor et al. 2003b). 1—Alluvium cover; Mesozoic: 2—dolerite and sediment; Paleozoic: 3—Amazon margin sediment; Neoproterozoic: 4—alkaline basalt (Cachoeira Seca), 5—alkaline plugs (Muri); Mesoproterozoic: 6—granite (Parguaza), 7—sediment (Vaupes supergroup), 8—basic sill/dyke (Avanavero), 9—sediment (Roraima Group), 10—acid plutono-volcanism (Uatumã); Transamazonian tectonothermal episode: 11—granitoid, 12—ultrabasic plug (Badidku), 13—greenstone belt, 14—granulite (Central Guiana); Archean: 15—granulite and migmatite (Imataca)

7.2 French Guiana

Geological mapping and mineral exploration in French Guiana from the late 1980ies onward (Marot 1988; Vanderhaeghe et al. 1998) resulted in an impressive series of publications by Claude Delor and his team in the journal *Géologie de la France* (Delor et al. 2003a, b). Based on extensive fieldwork, aeromagnetic, aeroradiometric mapping, structural data and zircon U-Pb, Pb-Pb and Sm-Nd geochronology, they published a new geological map of French Guiana, as well as a comprehensive plate-tectonics reconstruction of the whole northern part of the Guiana Shield (Fig. 7.5) as French Guiana is underlain almost exclusively by rocks of the Trans-Amazonian greenstone belt.

Delor et al. (2003a,b) subdivide the history of the Trans-Amazonian Orogeny in French Guiana and the eastern Guiana Shield in four stages, using the new official IUGS-approved subdivision of the Precambrian (Plumb 1991). It starts with a Eorhyacian period of ocean-floor spreading between earlier Amazonian and West-African plates (2.26–2.20 Ga), leading to a tholeiitic basaltic juvenile ocean crust evidenced in the lowermost parts of the greenstone belt, locally including also komatiites, and

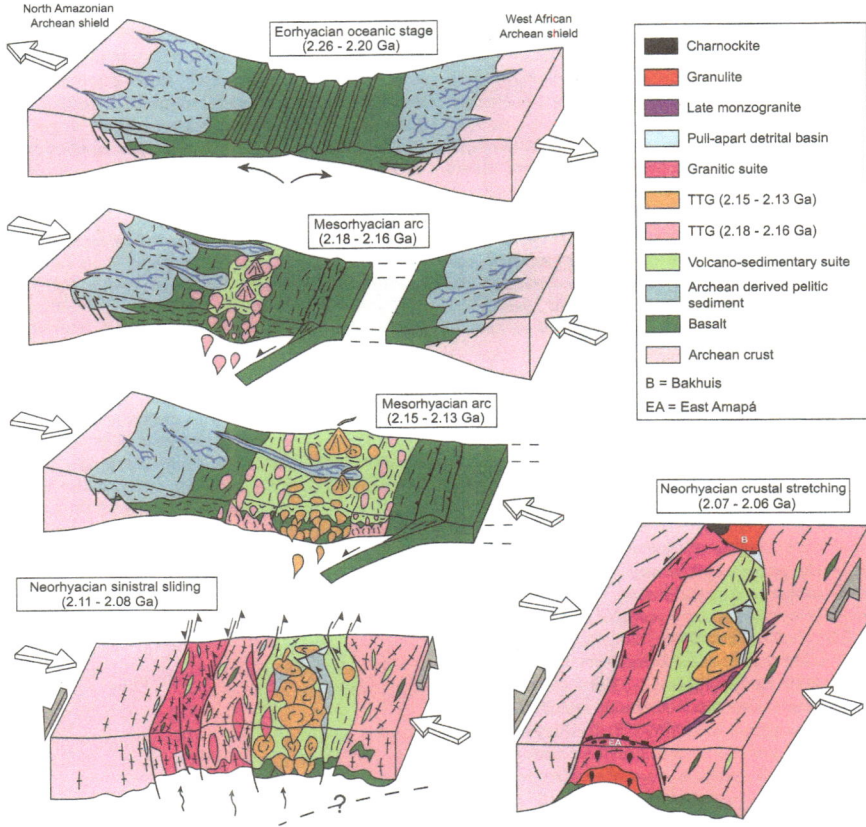

Fig. 7.5 Sequence of events in the greenstone belts of NE Guiana Shield (Delor et al. 2003a). Reproduced with permission from Géologie de la France

alternating with pyroclastics and chemical sediments. The oldest ages are represented by an U-Pb zircon age of 2208±12 Ma from an amphibolite at the Pointe des Amandiers in Cayenne (Delor et al. 2003a). In the Mesorhyacian between 2.18 and 2.13 Ga convergence between Amazonia and West-Africa starts in a southwards verging subduction zone, leading to their first stage of deformation D1, the intrusion of extensive sodic TTG plutons, andesitic to rhyolitic arc volcanism. They leave the question open whether this magmatism is the result of either Archean mantle upwelling leading to progressive oceanic crust melting or of subduction processes. The absence of HP-LT metamorphism would plead for the first hypothesis, but on account of geochemical arguments they prefer the latter. In the higher sections of the stratigraphy turbiditic greywackes appear, and on top of the greywackes epicontinental sediments were deposited in narrow troughs, mainly fluvial sandstones and conglomerates (Rosebel in Suriname, Orapu or Série Détritique in French Guiana).

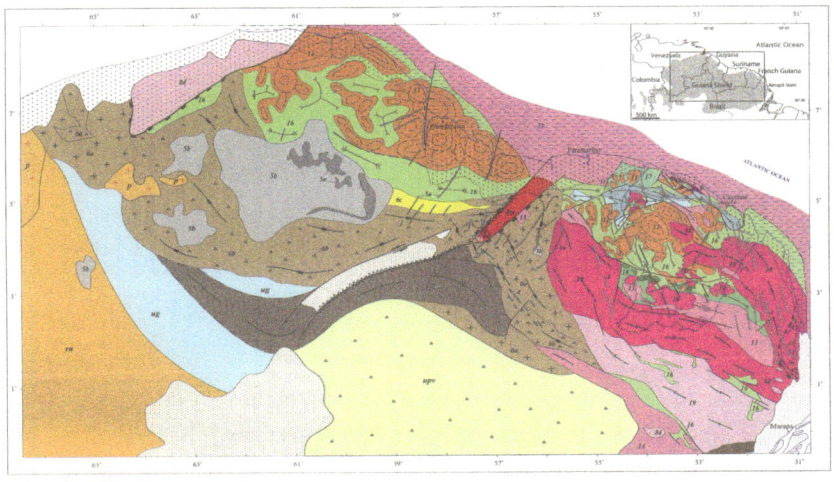

Fig. 7.6 Geological map of the Guiana Shield with its legend (Delor et al. 2003b). Reproduced with permission from Géologie de la France

The latter sediments are restricted to the eastern part of the greenstone belt in French Guiana and Suriname, east of the Bakhuis Granulite Belt. In the western part of the shield in Guyana these epicontinental sediments have not been recorded.

In the Neorhyacian between 2.11 and 2.08 Ga the deformation phase D2a was evidenced by sinistral sliding and metaluminous, more potassic granitoid magmatism. In stage 2b (2.07–2.06 Ga) opening and filling of pull-apart basins formed by crustal stretching, which ultimately suffered extensive high-grade metamorphism as in the Bakhuis granulite belt. Their views are illustrated by a series of cartoons that up to now are considered to render the best approximation of events during the Trans-Amazonian Orogeny (Figs. 7.5, 7.6). The same scheme is followed by Daoust (2016) and Daoust et al. (2011) in her paper about the Rosebel gold deposits in Suriname.

The large area occupied by felsic volcanic and shallow granites shallow south and west of the greenstone belt are depicted on Fig. 7.6 as *pro parte* Uatumã magmatism (now called Orocaima, see below) and they consider it a late stage of the Trans Amazonian Orogeny around 2.01–1.96 Ga, not depicted in the cartoons of Fig. 7.5.

7.3 Suriname

The Geological and Mining Service of Suriname was at that time in the process of producing a coloured 1:500,000 geological map of the country which appeared in 1977. The essence of their findings were published in a paper by Bosma et al.

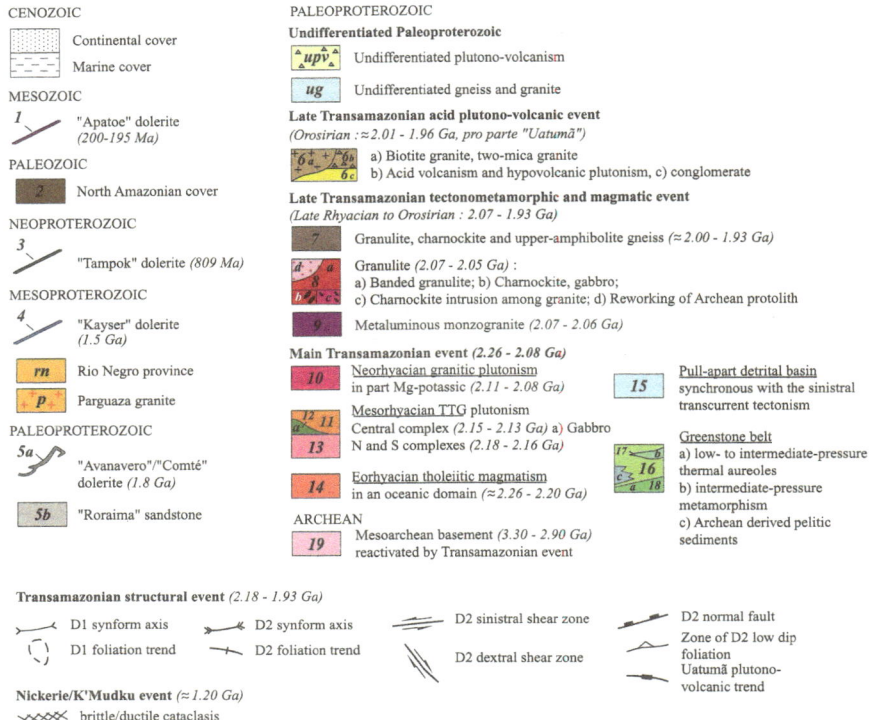

Fig. 7.6 (continued)

(1983), including a sketch map of the whole Guiana Shield (Fig. 7.7). An extensive explanation of the geological map of Suriname was published by Bosma et al. (1984), Rob de Vletter (1984) and further discussed by De Vletter et al. (1998). On this map, the area in western Suriname, occupied by the Central Amazonian Province of Brazilian accounts, is shown to be underlain by acid to intermediate volcanics and granitoid rocks.

Bosma et al. (1983) advocate an origin of the greenstone belt in Suriname along an active continental margin, starting with the extrusion of mafic, tholeiitic volcanics, followed by intermediate, calc-alkaline volcanic rocks, probably in a backarc marginal basin—island arc environment at the border of an older, pre-Trans-Amazonian continent. It was intruded by diapiric plutons, followed by the deposition of turbiditic sediments and later by coarse clastic epicontinental sediments. Finally large amounts of felsic magmatism, including ignimbritic volcanism, originated from greater depth in the subduction zone, representing almost complete reworking of the pre-Trans-Amazonian basement. The Rb-Sr isochron obtained by Priem et al. (1971) for Suriname basement rocks did not yet allow at that time to distinguish between the older diapiric tonalites in the greenstone belt and the younger Orocaima-type felsic magmatism. Gneiss and granulite belts probably originated at the same time. Maas

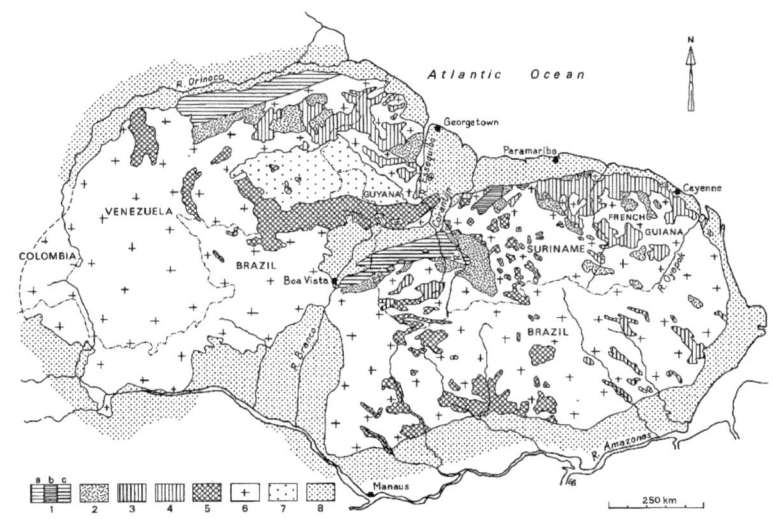

Fig. 7.7 Geological sketch map of the Guiana Shield. (Bosma et al. 1983)

(1979) designed a plate-tectonic scheme, in which the Trans-Amazonian Orogeny was portrayed as being the result of island arc formation, subduction and ultimately collision of an ancestral Guiana Craton with an Archean Imataca-Liberia Craton (Fig. 7.8).

In a subsequent review of the geology of the Guiana Shield, centered around Suriname (Fig. 1.3 in this text), Kroonenberg et al. (2016) subdivide the Trans-Amazonian units in three phases, which largely correspond to the subdivisions of Bosma et al. (1983) and Delor et al. (2003b) but have not been interpreted in specific deformation phases. The first phase is represented by a greenstone belt with ultramafic and mafic tholeiitic ocean floor lavas and intrusions, followed by island-arc andesites and dacites with intercalated volcanosedimentary rocks and chemical sediments. Since then detailed studies were made of the ultramafics (Naipal et al. 2019) and of manganese-rich chemical sediments (Amattaram et al. 2019) in the lowermost units. These units are intruded by TTG plutons, dated in French Guiana at 2.18–2.13 Ga (Delor et al. 2003a) and in Suriname between 2.18 and 2.12 Ga (Daoust 2016; Ramlal et al. 2019). Then follows a series of turbiditic metagreywackes and phyllites from which detrital zircon ages around 2.16 Ga have been found (Wijngaarde et al. 2019). They are intruded by S-type potassic granites with LCT-pegmatites in northeasternmost Suriname continuing into northwestern French Guiana, where they were dated at 2.08–2.06 Ga (Delor et al. 2003a) and in Suriname at 2.11 and 2.07 Ga (Kromopawiro et al. 2019). They produced an aureole of staurolite-garnet schists in the surrounding greywackes. A rhyolitic tuff layer in the overlying epicontinental conglomeratic-sandstone Rosebel Formation in Suriname was dated at 2.12 Ga (Ramlal et al. 2019).

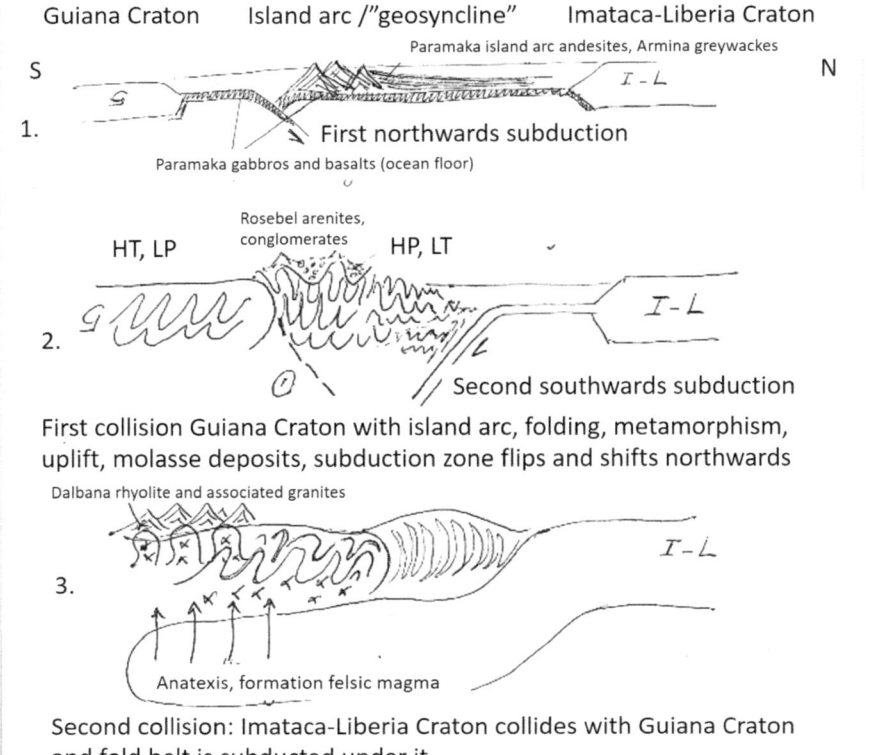

Guiana Craton   Island arc /"geosyncline"   Imataca-Liberia Craton

Fig. 7.8 'Actualistic' reconstruction of the Trans-Amazonian Orogeny (modified after Maas 1979)

The main synclinorium in the Suriname greenstone belt is bordered in both in the north and in the south by a zone of supracrustal migmatitic gneisses termed Sara's Lust Gneiss, (2.08 Ga, LA-ICPMS U-Pb rim, Kriegsman, unpublished). Whether they form an independent unit or should be considered as high-grade metamorphic greenstone-belt sediments in a transition to diatexitic granitoids further south remains to be investigated (Sastrohardjo et al. 2022).

According to Kroonenberg et al. (2016) a second Trans-Amazonian phase is represented by the high-grade metamorphic belts Bakhuis, Coeroeni, Kanuku and Cauarane, earlier designated as Central Guiana Granulite Belt (Kroonenberg 1976) with ages between 2.09 and 2.03 Ga (see Chapter 8). The shallow granites and felsic volcanics with ages around 1.99–1.96 Ga (now Orocaima), which include punctuated intrusions of coeval gabbros, charnockites and anorthosites, especially in the Bakhuis area, are considered to represent a third Trans-Amazonian phase (see also Chapter 8). The southern limit of the Trans-Amazonian orogeny in the Guiana Shield has been depicted by Kroonenberg et al (2019) to include the third-phase magmatism. They largely follow Delor et al. (2003a, b) in explaining the tectonic evolution of the Trans-Amazonian orogeny.

7.4 Amapá

In the Brazilian state of Amapá, the 'fourth Guiana' (once claimed by France), the greenstone belt continues eastwards. It was here, that McReath and Faraco (1996) obtained a Sm-Nd isochron age of 2.26 Ga for the Vila Nova greenstone belt, the oldest age for the Guiana Shield greenstone belts so far, but never repeated since then. At present, the only state-of-the art geochronological lab within the Guiana Shield, at the Universidade Federal do Pará (Ufpa) in Belém do Pará, is rapidly delivering important new age data from this area under the leadership of Jean-Michel Lafon and João Marinho Milhomem Neto. It was a major revelation when Valter de Avelar et al. (2003) and Lúcia Rosa-Costa et al. (2003) demonstrate that gneisses and granulites in a sizable part of the Amapá block have Archean protoliths. The main period of crust formation occurred during a protracted episode at the end of Paleoarchean and along the whole Mesoarchean (3.26–2.83 Ga). The protoliths were overprinted by Trans-Amazonian granulite-facies metamorphism between 2.10 and 2.09 Ga (Rosa-Costa et al. 2006, 2008) and between 2.06 and 2.04 Ga (Milhomem Neto et al. 2022). The Bacuri chromite-bearing mafic-ultramafic Bacuri complex in Amapá even represents the oldest rocks in the whole Guiana Shield with their U-Pb zircon age of 3343±3.5 Ma (Spier et al. 2022).

Lafon et al. (2019) state that In the Archean Amapá block, at least two episodes of crust generation are evidenced in the Eoarchean (~4.0 Ga) and the Mesoarchean (3.0–3.1 Ga). In the Rhyacian Lourenço and Carecuru domains, magmatic rocks are derived from mixed Rhyacian juvenile sources with Archean crustal components. In the Orosirian Erepecuru-Trombetas domain, magmatic rocks are originated from melting of mantle-derived magmas with participation of Rhyacian sialic crust and the existence of an Archean basement is discarded (see also chapter 8 and Fig. 10.2).

A small tip of the greenstone belt continues south of the Amazon Basin, the Bacajá domain (Vasquez et al. 2008; Macambira et al. 2009).

7.5 Venezuela

Also in Venezuela an impressive mapping and mineral exploration effort using side-looking radar, field geology, geochemistry and geophysics was carried out in their part of the Guiana Shield from 1987 to 1992 by the US Geological Survey and the Corporación Venezolana de Guayana (Wynn et al. 1995; Sidder and Mendoza 1995), ultimately resulting in a beautiful coloured geological relief map of Venezuela at scale 1:750,000 in two sheets (Hackley et al. 2005). They identified five lithotectonic provinces: (1) an Archean amphibolite- to granulite-facies gneiss terrane, (2) an Early Proterozoic greenstone-granite terrane(s), (3) an Early Proterozoic metamorphosed volcanic-plutonic complex, (4) Early to Middle Proterozoic continental sedimentary rocks, and (5) a Middle Proterozoic anorogenic rapikivi-type granite. The Archean age of the Imataca was based on 3.7–3.4 Ga whole-rock and feldspar U-Pb ages by

Montgomery (1979), and later confirmed by deformation, granite intrusion and meta-morphism around 2.8–2.7 Ga (Sidder and Mendoza 1995). The series was overprinted by Trans-Amazonian amphibolite-to granulite-facies metamorphism between 2.15 and 1.96 Ga. Later Rb-Sr, SHRIMP zircon U-Pb and Sm-Nd model ages by Tassi-nari et al. (2004) confirmed the>3.2 Ga protolith age, late Archean (~2.8 Ga) isotopic disturbance and juvenile accretion, and peak granulite-facies metamorphism between 2.05 and 1.98 Ga. The younger Pastora greenstone belt with Supamo TTG bodies give similar ages as elsewhere in the shield, around 2165 to 2080 Ga. According to Hildebrand et al. (2014) the Pastora group has been thrusted on top of the Imataca. Velázquez et al. (2011) consider the Pastora-Carichapo greenstone belts as having originated in an oceanic plateau.

References

Amattaram SM, Mason PRD, Kriegsman LM, Kroonenberg S (2019). Stratigraphy and Geochem-istry of the Paleoproterozoic manganese in the Guiana Shield, South America. Proceedings SAXI-XII Inter-Guiana Geological Conference 2022: Georgetown, Guyana 11–13

Avelar VG de, Lafon J-M, Delor C, Guerrot C, Lahondère D (2003) Archean crustal remnants in the easternmost part of the Guiana Shield: Pb-Pb and Sm-Nd geochronological evidence for Mesoarchean versus Neoarchean signatures. Géologie de la France 2–4:83–99

Bosma W, Kroonenberg SB, Maas K, de Roever EWF (1983) Igneous and metamorphic complexes of the Guiana shield in Suriname. Geol Mijnbouw 62:241–254

Bosma W, Kroonenberg SB, van Lissa R, Maas K, de Roever EWF (1984) Explanation to the geological map of Suriname 1:500,000. Meded Geol MijnbKundige Dienst Suriname 27:31–82

Daoust C (2016) Caractérisation stratigraphique, structurale et géochimique du District minéralisé de Rosebel (Suriname) dans le Cadre de l'évolution géodynamique du Bouclier Guyanais, PhD thesis, Université du Québec á Montréal (Montréal): 330

Daoust C, Voicu G, Brisson H, Gauthier M (2011) Geological setting of the Paleoproterozoic Rosebel gold district, Guiana Shield, Suriname. J S Am Earth Sci 32:222–245

Delor C, Lahondère D, Egal E, Lafon JM, Cocherie A, Guerrot C, de Avelar V (2003a) Transamazo-nian crustal growth and reworking as revealed by the 1:500,000-scale geological map of French Guiana. Géologie de la France 2–4:5–57

Delor C, de Roever EWF, Lafon JM, Lahondère D, Rossi Ph, Cocherie A, Guerrot C, Potrel A (2003b) The Bakhuis ultra-high temperature granulite belt Suriname : II implications for late Trans-Amazonian crustal stretching in a revised Guiana Shield framework. Géologie de la France 2–4: 207–230

De Vletter DR (1984) Synthesis of the Precambrian of Suriname and review of some outstanding problems. In: De Vletter DR (eds) Mededelingen Geologisch Mijnbouwkundige Dienst Suriname 27:11–30

De Vletter DR, Aleva GJJ, Kroonenberg SB (1998) Research into the Precambrian of Suriname. In: Wong et al (eds), The history of earth sciences in Suriname Royal Netherlands Academy of Sciences/Netherlands Institute of Applied Geosciences 15–64

Gibbs AK (1980) Geology of the Barama-Mazaruni Supergroup of Guyana. Ph.D. thesis, Harvard University, Cambridge, Mass., USA 385

Gibbs AK, Olszewski WJ (1982) Zircon U-Pb ages of Guyana greenstone-gneiss terrane. Precambr Res 17:199–214

Gibbs AK, Barron CN (1983) The Guiana Shield reviewed. Episodes 7–14

Gibbs AK, Barron CN (1993) Geology of the Guiana shield. Oxford University Press 246

Hackley PC, Urbani F, Karlsen AW, Garrity, CP (2005) Geologic shaded relief map of Venezuela. USGS Open File Report 2005–1038

Hildebrand RS, Buchwaldt R, Bowring SA (2014) On the allochthonous nature of auriferous greenstones, Guayana shield, Venezuela. Gondwana Res 26:1129–1140

Kromopawiro S, Kroonenberg SB, Kriegsman LM, Mason PRD (2019) 2.12–2.08 Ga Late- to post-collisional peraluminous granitoid magmatism in the Marowijne Greenstone Belt of Suriname. In Proceedings 11th Inter Guiana Geological Conference, Paramaribo. Mededeling Geologisch Mijnbouwkundige Dienst Suriname 29:105–109

Kroonenberg SB (1976) Amphibolite-facies and granulite-facies metamorphism in the Coeroeni-Lucie área, SW Surinam. Thesis Amsterdam, Mededelingen Geologisch Mijnbouwkundige Dienst Suriname 25:109–289

Kroonenberg SB, de Roever EWF, Fraga LM, Reis NJ, Faraco MT, Cordani UG, Lafon J-M, Wong TE (2016) Paleoproterozoic evolution of the Guiana Shield in Suriname—a revised model. Neth J Geosci-Geol En Mijnb 95:491–522

Kroonenberg S, Mason PRD, Kriegsman LM, de Roever EWF, Wong TE (2019) Geology and mineral deposits of the Guiana Shield. In: Proceedings 11th Inter Guiana Geological Conference, Paramaribo. Mededeling Geologisch Mijnbouwkundige Dienst Suriname 29:111–115

Lafon JM, Rosa-Costa L, Milhomem Neto JM (2019) Sr-Nd-Hf isotopic tracing of Archean continental crust in the Brazilian part of the Southeastern Guyana Shield: A review. In: Proceedings 11th Interguiana Geological Conference 2019: Paramaribo, Suriname. Mededeling Geologisch Mijnbouwkundige Dienst Suriname 29:121–124

Maas K (1979) Nota betreffende een overzicht, alsmede een tentatieve interpretatie van het Precambrium van Suriname. Geologisch Mijnbouwkundige Dienst Suriname, internal report 23

Macambira MJB, Lacerda Vasquez M, Costa da Silva DC, Galarza MA, de Mesquita Barros CE, de Freitas CJ (2009) Crustal growth of the central-eastern Paleoproterozoic domain, SW Amazonian craton: Juvenile accretion versus reworking. J S Am Earth Sci 27:235–246

Marot A (1988) Carte Géologique de la France à 1/500 000 Guyane Sud, Note explicative. Orléans 86

McReach I, Faraco MT (1996) Sm-Nd and Rb-Sr systems in part of the Vila Nova metamorphic suite, northern Brazil, Extended abstract South American Symposium on Isotope Geology, São Paulo 194–196

Milhomem Neto JM, Lafon JM, de P. Amaral Ferreira D, Silva Miranda S, Dantas EL (2022) High-grade metamorphism in the central region of Amapá, Northern Brazil: age constraints from in situ U-Pb dating of monazite and zircon. SAXI-XII Inter-Guiana Geological Conference 2022: Georgetown, Guyana, pp 105–10

Montgomery CW (1979). Uranium-Lead Geochronology of the Archean Imataca Series, Venezuelan Guayana Shield. Contrib. Mineral. Petrol. 69:167–176

Naipal R, Kroonenberg S, Mason PRD (2019) Ultramafic rocks of the Paleoproterozoic greenstone belt in the Guiana Shield of Suriname, and their mineral potential. Proceedings 11th Inter Guiana Geological Conference, Mededeling Geologisch Mijnbouwkundige Dienst Suriname 29:143–146

Norcross C, Davis DW, Spooner ETC, Rust A (2000). U-Pb and Pb-Pb age constraints on Paleoproterozoic magmatism, deformation and gold mineralization in the Omai area, Guyana Shield. Precambrian Research 102:69–86

Plumb KA (1991) New Precambrian time scale. Episodes 14:139–140

Priem HNA, Boelrijk NAIM, Hebeda EH, Verdurmen EAT, Verschure RH (1971) Isotopic ages of the Trans-Amazonian acidic magmatism and the Nickerie Episode in the Precambrian basement of Surinam, South America. Geol Soc Am Bull 82:1667–1680

Ramlal S, Kroonenberg SB, Mason PRD, Kriegsman LM, O'Sullivan P (2019) Multiphase TTG intrusions in the Paleoproterozoic greenstone belt of Suriname and their role in gold mineralization in the Rosebel gold district. In: Proceedings 11th Inter Guiana Geological Conference, Paramaribo. Mededeling Geologisch Mijnbouwkundige Dienst Suriname 29:159–162

Rosa-Costa LT, Ricci PSF, Lafon J-M, Vasquez ML, Carvalho JMA, Klein EL, Macambira EMB (2003) Geology and geochronology of Archean and Paleoproterozoic domains of southwestern Amapá and northwestern Pará, Brazil, southeastern Guiana shield. Géologie de la France, 2003, 2–3–4:101–120

Rosa-Costa LT, Lafon J-M, Delor C (2006) Zircon geochronology and Sm–Nd isotopic study: Further constraints for the Archean and Paleoproterozoic geodynamical evolution of the southeastern Guiana Shield, north of Amazonian Craton, Brazil. Gondwana Res 10:277–300

Rosa-Costa LT, Lafon J-M, Cocherie A, Delor C (2008) Electron microprobe U-Th-Pb monazite dating of the Transamazonian metamorphic overprint on Archean rocks from the Amapá Block, southeastern Guiana Shield, Northern Brazil. J S Am Earth Sci 26:445–462

Sastrohardjo F, Vanderhaeghe O, Kriegsman L, Eglinger A, Kroonenberg S, Bardoux M (2022) Nature of the relationship between the Marowijne greenstone belt and the Gran Rio granite of the Rhyacian Transamazonian orogenic belt, Suriname: Significance of the Sara's Lust migmatite. In Proceedings SAXI- XII Inter-Guiana Geological Conference 2022: Georgetown, Guyana, pp 128–134

Sidder GB, Mendoza V (1995) Geology of the Venezuelan Guayana Shield and its relation to the geology of the entire Guayana Shield. In: Sidder et al. (eds) Geology and Mineral Deposits of the Venezuelan Guiana Shield. U.S. Geological Survey Bulletin 2124, B1-B41

Snelling NJ (1995) Book review: The geology of the Guiana Shield. J S Amn Earth Sci 8:123–125

Spier CA, Ferreira Filho CF, Daczko N (2022) Zircon U-Pb isotopic and geochemical study of metanorites from the chromite-mineralised Bacuri Mafic-Ultramafic Complex: Insights of a Paleoarchean crust in the Amapá Block, Guyana Shield, Brazil. Gondwana Res 105:262–289

Tassinari CCG (1981) Evolução geotectônica da Província rio Negro-Juruena na região amazônica. Instituto de Geociências, Universidade de São Paulo, Dissertação de mestrado, p 99

Tassinari CCG, Munhá JMU, Teixeira W, Palacios T, Nutman A, Sosa SC, Santos AP, Calado BO (2004) The Imataca Complex, NW Amazonian Craton, Venezuela: crustal evolution and integration of geochronological and petrological cooling histories. Episodes 27:3–12

Tedeschi MT, Hagemann S, Kemp AIS, Kirkland CL, Ireland TR (2020) Geochronological constrains on the timing of magmatism, deformation and mineralization at the Karouni orogenic gold deposit: Guyana, South America. Precambr Res 337:105329

Vanderhaeghe O, Ledru P, Thiéblemont D, Egal E, Cocherie A, Tegyey M, Milési JP (1998) Contrasting mechanism of crustal growth. Geodynamic evolution of the Paleoproterozoic granite–greenstone belts of French Guiana. Precambr Res 92:165–193

Vasquez ML, Macambira MJB, Armstrong RA (2008) Zircon geochronology of Granitoids from the western Bacajá domain, southeastern Amazonian craton, Brazil: Neoarchean to Orosirian evolution. Precambr Res 161:279–302

Velázquez GD, Béziat S, Salvi T, Tosiani DP (2011) First occurrence of Paleoproterozoic oceanic plateau in the Guiana Shield: The gold-bearing El Callao Formation, Venezuela. Precambr Res 186:181–192

Wijngaarde GW, Kroonenberg SB, Mason PRD, Kriegsman LM (2019) Petrography, geochemistry and age of the Armina Formation metaturbidites of the Coppename River, Suriname. In: Proceedings 11th Inter Guiana Geological Conference, Paramaribo. Mededeling Geologisch Mijnbouwkundige Dienst Suriname 29, pp 201–205

Wynn JC, Sidder GB, Gray G, Page NJ, Mendoza V (1995) The cooperative project between the U.S. Geological survey and the corporacion Venezolana de Guayana, Tecnica Minera, C.A, in the Venezuelan Guayana Shield, Estado Bolivar and Estado Amazonas, Venezuela. In: Sidder et al (eds) Geology and Mineral Deposits of the Venezuelan Guiana Shield. U.S. Geological Survey Bulletin 2124 A1–A7

Chapter 8
Challenging the Brazilian Geochronological Provinces: The Fraga Model

Abstract The model of continental growth by accretion of younger geotectonic provinces around an Archean nucleus was challenged by the discovery of sinuous high-grade belts in the central Guiana Shield that crossed the geochronological provinces. In fact it demonstrated it is just as important to take into account lithostratigraphy and structure in the tectonic history, as the age of primary crystallisation of magmas from the mantle on which the geochronological provinces are based. Only the Rio Negro Block in the westernmost part of the Guiana Shield accreted during the Querary Orogeny (1.85-1.72 Ga). Mantle plumes are held responsible for the extensive felsic Orocaima magmatism (1.99-1.96 Ga) in the central Guiana Shield and the 1.89-1.82 Ga Uatumã felsic magmatism in the southernmost part.

8.1 The Cauarane-Coeroeni Belt

In 2009 the Brazilian CPRM geologist Lêda Fraga published a paper on 1.94–1.93 Ga charnockitic magmatism in Roraima State in Brazil (Fraga et al. 2009b). In it she presented a map of the central Guiana Shield in which the Brazilian standard subdivision in NNW-SSE accretionary belts around the Central Amazonian Province was challenged. In a way, she restored the primacy of lithostratigraphy and structure over the reconstruction of continental growth. She portrayed a sinuous east-west running high-grade belt straight across Cordani's and Tassinari's provinces which connected the Cauarane gneiss belt in Roraima state through the Kanuku high-grade belt in Guyana with the Coeroeni Gneiss belt in southwestern Suriname, and called it Cauarane-Coeroeni Belt (CCB). She was inspired by the older concept of the Central Guiana Granulite Belt, which also connected Coeroeni with the Cauarane area in Roraima state (Kroonenberg 1976). In later papers she elaborated the concept in more detail, supported by many U-Pb zircon age data (Fraga et al. 2017, Fraga et al. 2009a, 2024; Figs. 8.1, 8.2).

Furthermore she portrays an equally sinuous E-W running belt north of the CCB of around 200,000 km^2 between the Orinoco River and western Suriname, consisting of shallow granitoids and felsic volcanics, with ages between 1.99 and 1.96 Ga. She

S. Kroonenberg, *The Changing Framework of the Guiana Shield*, SpringerBriefs in Earth System Sciences, https://doi.org/10.1007/978-3-031-86334-9_8

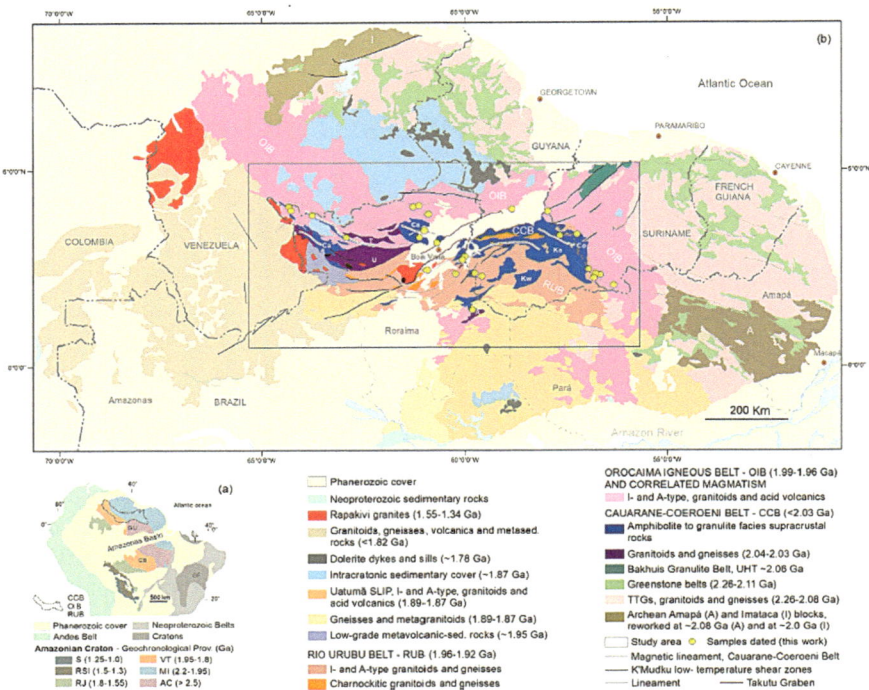

Fig. 8.1 Geological map of the Guiana Shield according to Fraga et al. (2024)

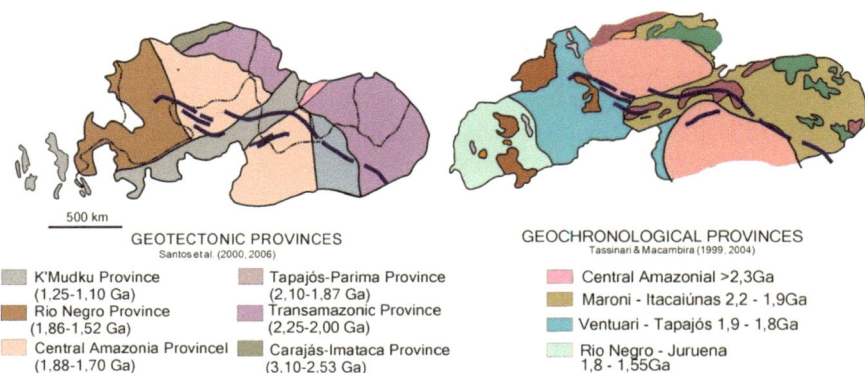

Fig. 8.2 Fraga's Cauarane-Coeroeni belt (in blue lines) runs across postulated accretionary belts in the central Guiana Shield (Fraga 2009, redrawn)

designates it as Orocaima Igneous Belt, following Reis et al. (2000, 2021, see par. 8.2). At last she postulated a younger sinuous E-W running Rio Urubu Belt (RUB) of I and A-type granitoids, gneisses and granitoids with ages between 1.96 and 1.92 Ga south of the CCB. Also these units cross the boundaries of the geochrono-logical boundaries of Cordani et al. (1979), Teixeira et al. (1989) and Tassinari and Macambira (1999). Fraga's three main belts deserve some more detailed discussion.

Fraga interpreted the Cauarane-Coeroeni belt as a collisional belt between an older Rhyacian, Trans-Amazonian plate and the 2.04–2.03 Ga Trairão and Anauá magmatic arcs in a northwards dipping subduction zone (Fraga et al. 2017, 2019, 2024). The metamorphism in the Cauarane-Coeroeni Belt took place in two phases, according to U-Pb zircon data by Fraga et al. (2024). The age of the first, highest-grade and most pervasive phase, related to continental collision could not be resolved, but was considered to be <2.03 Ga. A minor second phase around 1.98 Ga is represented as narrow rims around the zircons. The great majority of older zircon grains is thought to be of detrital origin. According to her out of 360 spots in these detrital zircons with less than 10% of discordance 257 grains indicate a derivation from Paleoproterozoic rocks and 103 record Archean sources.

However, it appears from her Table 8 and Supplementary Material 2 that at least 140 of those supposedly detrital Paleoproterozoic zircons show ages between 2.09 and 2.03 Ga, ages that are mostly too young to have been derived from the 2.26–2.08 Ga Rhyacian greenstone belt. This leads to the suspicion that part of these zircons are not of detrital origin but reflect the first high-grade phase of metamorphism, in spite of their Th/U ratios >0.1. Yakymchuk et al. (2018) show that higher Th/U ratios in high-grade metamorphic rocks are not uncommon.

Metamorphic ages between 2.09 and 2.03 Ga are well-known from the supracrustal granulites in the Bakhuis UHT Granulite Belt in western Suriname, partly also with Th/U ratios >0.1 (De Roever et al. 2003b, 2019, 2022) and Nanne et al. (2020). In older literature this belt was often suggested to be of Archean age because of the pervasive granulite facies metamorphism as in the Imataca block (see chapter 3). However, all U-Pb zircon ages obtained from the granulites point to a Rhyacian age (De Roever et al. 2003b, 2019, 2022), and so the question arises whether the Bakhuis belt could form part of the Cauarane-Coeroeni Belt as well. They share a supracrustal origin and an anticlockwise metamorphic history, evidenced espe-cially by the replacement of cordierite by higher-pressure minerals in both belts, which pleads against a collisional origin (De Roever et al. 2003b; Delor et al. 2003b, Kroonenberg et al. 2016). This hypothesis is corroborated by the occurrence of orthopyroxene-bearing pelitic granulites, a typical Bakhuis assemblage (De Roever et al. 2003b) in the westwards continuation of the Bakhuis Belt across the Corantijn River into Guyana towards the CCB (Kroonenberg 1975). However, in aeromagnetic images this connection is not very evident, as the pattern is strongly influenced by the SW-NE K'Mudku/Nickerie shear zones and the Takutu rift valley. Nevertheless, we consider Bakhuis and CCB together as a kind of triple junction, originated by sedimentation in a rift-type setting and high-grade metamorphism in a second phase of the Trans-Amazonian Orogeny between 2.09 and 2.03 Ga by southward subduction and finally continental collision between a West-African Craton and an ancestral Guiana Shield.

In previous papers from other authors various opinions on the relation between these belts are expressed: Gibbs and Barron (1983, 1993, (here Fig. 7.1 and 7.3), Teixeira et al. (1989; Fig. 6.4), Tassinari and Macambira (1999; Fig. 6.5) follow the triple junction pattern, but the same Gibbs and Barron in their 1993 coloured map (Fig. 7.4) draw Bakhuis continuing into the Kanuku belt of Guyana, separating Coeroeni as an individual unit. Delor et al. (2003), Klaver et al. (2015) and Reis et al. (2021, Fig. 8.4) follow Fraga et al. (2024, Fig. 8.1), by considering the Bakhuis UHT granulite belt as a separate unit unconnected to other high-grade belts.

A related complication is the fact that Santos (2003) and Santos et al. (2000, 2006a, b), incorporate the Bakhuis Granulite Belt in a so-called K'Mudku Shear belt (Chapter 6.3, Figs. 6.8 and 6.11). Now as explained before (Chapter 6.3), the K'mudku Mylonitic Episode event was coined by Barron (1969) as a ENE-WSW shear zone in southern Guyana, which showed low-grade thermal resetting of mica ages around 1200 Ma, but not as a collisional orogenic belt. Moreover, thermal mica rejuvenating is not only evidenced in mylonite zones, but also in non-cataclastic rocks in the whole of the western shield: in Suriname (Priem et al. 1968a, 1971, 1977), in Venezuela (Martín Bellizzia 1972) in Colombia (Priem et al. 1982). It is an overprint over older, mainly Trans-Amazonian rocks and should not have been marked as a separate province on Santos' maps. It has been ascribed to the Grenvillian collision between Laurentia and Amazonia (1.2–1.0 Ga, Kroonenberg 1982, 2019; Kroonenberg and De Roever 2010; Cordani et al. 2010; Ibáñez-Mejía et al. 2011; Fig. 8.3).

8.2   The Orocaima and Urubu Belts

The second belt discussed by Fraga et al. (2024) is the series of 1.99–1.96 Ga felsic metavolcanics and shallow granitoids (Fig. 8.2), originally coined as Orocaima by Reis et al. (2000). Soon after Reis' publication it became clear that there are actually *two* provinces with felsic volcanics and shallow granitoids. At present the name Orocaima is only retained for the older series. The Orocaima belt presents ages between 1.99 and 1.96 Ga and cover at least 200,000 km^2 in western Suriname, southern Guyana, Venezuela and northern Roraima Province in Brazil (Caicara, Surumu, Dalbana volcanics, Cuchivero, Pedra Pintada and Wonotobo granitoids, Iwokrama in Guyana for both volcanics and granitoids; Reis et al. (2003), Anandbahadoer-Mahabier and de Roever 2019; Barbosa et al. 2021). Even along the Orinoco River Caicara felsic volcanics (Kroonenberg 2019) and Cuchivero granitoids (Ibáñez-Mejía and Cordani 2020) have been recorded, extending its distribution westwards to at least the frontier with Colombia. A younger province (Iricoumé volcanics and associated Mapuera and Madeira granites) in the Uatumã River area in the southernmost Brazilian part of the Guiana Shield shows

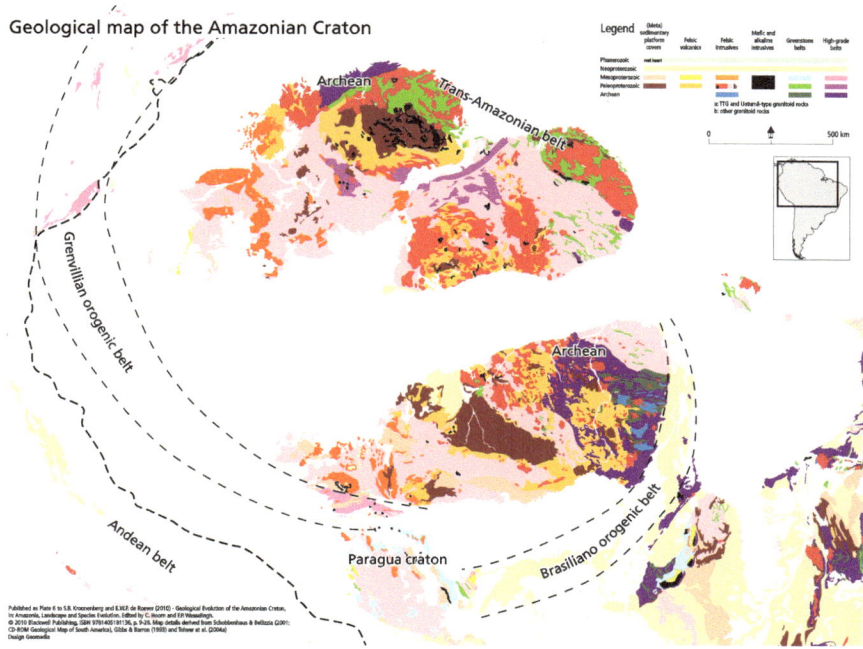

Fig. 8.3 Geological map of the Amazonian Craton, also showing the Andean Grenvillian granulite belt and its counterparts in the southwestern craton. The Guiana Shield part is since then superseded by Fig. 1.3. (from Kroonenberg and De Roever 2010; Reproduced with permission Wiley-Blackwell)

ages around 1.89–1.82 Ga (Reis et al. 2021; chapter 8.3). The younger one retains the name Uatumã belt, which is unfortunate as also this name was used in earlier publications for both provinces, before their differing ages were known (Ramgrab and Santos 1974; Montalvão 1975; Gibbs and Barron 1993; Delor et al. 2003b). Reis et al. (2021) consider both provinces as SLIPs (Siliceous Large Igneous Provinces).

Verhofstad (1971) was one of the first to establish the ignimbritic character of the volcanics in the Orocaima belt. Barbosa et al. (2021) give an extensive description of the Orocaima SLIP ignimbritic felsic volcanics in northern Brazil, whose volcanic breccias rich in lithics and lithic lapilli-tuff facies show the proximity of the source. However, the Orocaima event is more complex than just felsic magmatism. It also contains a series of punctuated (ultra)mafic intrusions with calcalkaline affinities, called Appinitic Suite in southern Guyana by Berrangé (1977) and Lucie Gabbro in Suriname by Kroonenberg et al. (2016), who report a Pb-Pb evaporation age of 1985 Ma from the type locality, which later was updated with a U-Pb zircon SIMS age of 1996 Ma (unpublished). Furthermore, within the UHT Bakhuis Granulite belt a layered anorthosite pluton gave a Pb-Pb evaporation age of 1980 Ma (De Roever et al. 2003), several charnockite plutons show ages between 1.98 and 1.99 Ga (Klaver

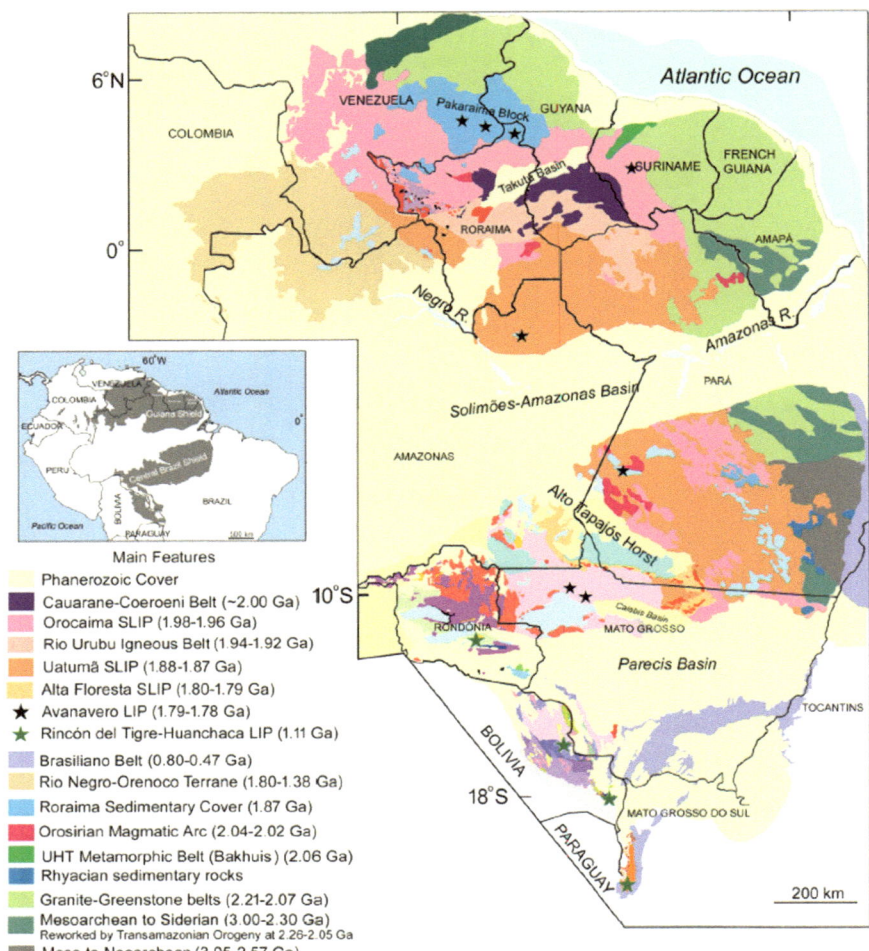

Fig. 8.4 Orocaima and Uatumã SLIPs in the Amazonian Craton (from Reis et al. 2021, Fig. 1, reproduced with permission from GSL)

et al. 2015), and two gabbroic plutons in the Bakhuis Granulite Belt were dated at 1.98 Ga (U-Pb zircon) and 1.97 Ga (baddeleyite), respectively (Klaver et al. 2016).

The Orocaima event also brought about some metamorphic effects. This is obvious from the fact that numerous zircons from the Cauarane-Coeroeni Belt have metamorphic rims between 1999 and 1933 Ma with Th/U ratios <0.1 (Fraga et al. 2024). This is the static metamorphic phase M2 of Fraga et al. (2024), responsible for the replacement of low-pressure mineral assemblages with cordierite by higher pressure ones: the second limb of the anticlockwise cooling path of the Cauarane-Coeroeni-Bakhuis rocks (De Roever et al. 2003b; Kroonenberg et al. 2016). Klaver et al. (2015)

consider the 1.99–1.98 Ga charnockites in the Bakhuis Granulite belt as evidence for a second UHT event. Furthermore the Werekitto gneisses, formerly thought to represent quartzofeldspathic gneisses belonging to the Rhyacian Coeroeni Group (Kroonenberg et al. 2016), now show SHRIMP U–Pb zircon ages of 1990–1985 Ma (Fraga et al. 2024).

The Orocaima felsic volcanics in northwestern Suriname and eastern Guyana are conformably underlain by gently folded quartzarenites and quartz conglomerates of the Ston Formation (Suriname) and the Muruwa Formation (Guyana) (Loemban Tobing 1969, Bosma et al. 1983). Detrital zircons from the Ston formation show two LA-ICPMS U-Pb age populations: one around 2.11 Ga, which obviously represents a greenstone provenance, and one around 1.98 Ga (Kroonenberg, unpublished data). The latter age suggests that the Ston Formation was still being deposited when the Orocaima volcanism started. The Ston Formation shows a constant southwestwards tectonic dip indicating syn-or post-Orocaima deformation.

At last, recently a radiating swarm of 1.98 Ga tholeiitic dolerite dykes was discovered in northern Venezuela and northern Suriname, and interpreted as a mafic Large Igneous Province. The radial pattern of the dykes and their age suggest a mantle plume triggered the Orocaima SLIP (Ibáñez-Mejía et al. 2020).

Fraga et al. (2024) consider the 1.99–1.96 Ga Orocaima Igneous Belt as having originated in a late-orogenic, post-collisional setting related to an Early Orosirian continental collision in a late stage of the Trans-Amazonian Orogeny that gave rise to the Cauarane-Coeroeni Belt (see also discussion by Barbosa et al. 2021). This stage was denominated Akawai Orogeny by Fraga and Cordani (2019) and Fraga et al. (2024), thus persisting in the misnomer already introduced by Issler (1975, see chapter 5). Kroonenberg et al. (2016) did not recognise a separate 'Akawai' orogeny, but consider that on the basis of the position of this belt parallel to the greenstone belt, its calc-alkaline nature and syntectonic collision trace-element geochemistry, the felsic volcanic-granitoid complex represents the third stage of the Trans-Amazonian Orogeny. This may be attributed to underplating of large quantities of mafic magma and as a corollary extensive crustal melting. The persistent occurrence of small (ultra) mafic plutons associated with the Orocaima volcanics and granites may testify to that origin.

The granitoids and gneisses of the Rio Urubu Belt, the third belt in Fraga's (2024) scheme, show U-Pb zircon ages between 1973 and 1935 Ma, averaging 1949 Ma disregarding grains >2000 Ma considered inherited according to the Supplementary data. Orocaima zircons average 1977 Ma, without taking into account inherited and disregarded zircons. These ages overlap with younger 1974–1949 Ma ages from Orocaima rocks e.g. in Suriname (Kroonenberg et al. 2016). Gneisses are not uncommon in the Orocaima Igneous belt either, as shown by the Werekitto gneisses mentioned above. Because of the limited amount of age data from the Urubu Belt its relation with the CCB and Orocaima remains to be established.

8.3   The Uatumã Belt

The Uatumã belt in its present SLIP significance (Fig. 8.4) also consists of felsic volcanics (Iricoumé in Brazil, Kuyuwini in Guyana) and shallow granitoids (Mapuera, Madeira in Brazil), and its individuality was only recognised when dating showed it to be younger than the Orocaima belt, around 1.89–1.82 Ga (Klein et al. 2012; Reis et al. 2021). The youngest ages around 1.82 Ga are represented by the Sn-cryolite Madeira caldera complex in the Pitinga district (Bastos Neto et al. 2009). It continues across the Amazon basin into the Iriri-Xingu area of the Guaporé Shield and covers there large surfaces.

Now this 1.89–1.82 Ga Uatumã belt coincides exactly with the area indicated by Cordani et al. (1979) and Tassinari and Macambira (1999) as the Guiana Shield part of the Central Amazonian Province around which all other geochronological provinces would have been accreted. As for the Uatumã unit in the Guiana Shield, Lafon et al. (2019) conclude on the base of Sm-Nd systematics and the lack of inherited Archean zircons, that the Archean Central Amazonian Province should not be extended to the Guiana Shield. These data, together with those by Fraga et al. (2024) cited above, suggest that there is as yet no evidence for continental accretion around an Archean Central Amazonian Province in the Guiana Shield.

8.4   The Río Negro and Younger Belts

The only area for which the accretion scenario according to Tassinari and Macambira (1999) is maintained is the Rio Negro block in the westernmost part of the Guiana Shield. The position of the suture zone with the older province further east is not firmly established (Ibáñez-Mejía and Cordani 2020), but according to recent aeromagnetic and gravity data (Moyano-Nieto and Prieto. 2021) it runs approximately N-S in Venezuela east of the Orinoco River, unfortunately in an area from which few modern age data are available (undivided Proterozoic rocks, Sidder and Mendoza 1995; Hackley et al. 2005). The Rio Negro Block is underlain by the migmatitic gneisses of the Mitú complex in Colombia (Galvis et al. 1979; Huguett et al. 1979; Priem et al. 1982), or Cauaburí gneisses (Brazil) dated between 1.85 and 1.72 Ga, and intruded by at least two suites of granite plutons around 1.72 and 1.55 Ga (Almeida et al. 2013, 2020, 2022, 2023; Cordani et al. 2016; Kroonenberg 2019; Bonilla-Pérez et al. 2021). The Parguaza rapakivi granite that straddles the Venezuelan-Colombian border (Bonilla-Pérez 2013) forms part of a discontinuous NW-SE stretching belt of ~1.55 Ga anorogenic rapakivi granites that extends far east to the Mucajaí granite in Roraima state (Heinonen et al. 2012). The Matraca rapakivi granite of 1343 Ma along the Inírida River in eastern Colombia is a younger representative of that series (Bonilla-Pérez et al. 2016).

Also within this Rio Negro block there is a diversity of opinions about its internal structure, partly because of its extensive sedimentary cover. Almeida et al. (2013,

2022, 2023) distinguished two orogenies in the high-grade metamorphic complex in the Brazilian part Río Negro block: the Cauaburí Orogeny from 1.81–1.75 Ga and the Querarí Orogeny (1.74–1.70 Ga) and a third one, the Içana Orogeny for the 1.55–1.50 granitoid intrusions. Kroonenberg (2019) considers the high-grade metamorphic event as a single orogenic event, retains only the name Querarí Orogeny and considers the granites as anorogenic. Cordani et al. (2016) distinguished, faithful to his continental growth models advocated since 1979, two NW-SE stretching belts in the block, a first orogenic pulse in the easternmost Atabapo Belt 1.80–1.74 Ga, and a second orogenic pulse in the Vaupés belt 1.58–1.50 Ga. Ibáñez-Mejía and Cordani (2020), however, state that it remains unclear whether these regional magmatic episodes are indeed representative of distinct crust-forming events that affected two separate geographic regions, or if the entire region consists essentially of Paleoproterozoic basement that was later affected by Mesoproterozoic post–tectonic or anorogenic magmatism. U-Pb zircon crystallisation ages seem to support the latter point of view (Ibáñez-Mejía and Cordani 2020). The moderately positive to negative εHf and εNd values at time of crystallization obtained from almost all mid Paleo– to early–Mesoproterozoic intrusives in the westernmost Guiana Shield indicate the widespread incorporation of some proportion of reworked older crustal material with Rhyacian (mainly Trans-Amazonian) T_{DM} ages (Ibáñez-Mejía and Cordani 2020).

Recently evidence for a still younger province at its western extremity has been published: the Guaviare Complex in Colombia, of 1.3 Ga (Amaya-López et al. 2020). Its significance still has to be elaborated.

The youngest accretionary event in the Precambrian is the Grenvillian or Putumayo orogeny evidenced by granulite-facies metamorphism and calc-alkaline magmatism recorded in the Colombian Andes and in drill cores in the Andean Foredeep (Kroonenberg 1982, 2019; Ibáñez-Mejía et al. 2011). Its impact on the Guiana Shield is marked by extensive shearing and mylonitization, low-grade thermal resetting of mineral ages (the K'Mudku/Nickerie event, see chapter 6.3 and 8.1) and minor alkaline magmatism (Cordani et al. 2010).

References

Almeida ME, Macambira MJB, Santos JOS, do Nascimento RSC, Paquette J-L (2013) Evolução crustal do noroeste do Cráton Amazônico Amazonas, Brasil baseada em dados de campo, geoquímicos e geocronológicos. Anais do 13º Simpósio de Geologia da Amazônia 201–204

Almeida ME, Nascimento R (2020) Geologia e evolução crustal do centro-norte do Cráton Amazônico e correlações com as províncias geocronológicas. In: Bartorelli A, Teixeira W, de Brito Neves B.B (eds), Geocronologia e Evoluçao Tectônica do Continente Sul-Americano: a contribuição de Umberto Giuseppe Cordani 111–122

Almeida ME, Nascimento RSC, Mendes TA, Santos TA, Macambira MJB, Vasconcelos P, Pinheiro SS (2022) An outline of Paleoproterozoic-Mesoproterozoic crustal evolution of the NW Amazon craton and implications for the Columbia Supercontinent. Int Geol Rev. https://doi.org/10.1080/00206814.2021.2025158

Almeida ME, Macambira MJB, Mendes TAA, Santos JOS (2023) Post-collisional 1.75 Ga mantle-derived felsic magmatism in the Cauaburi Orogeny, NW Amazon Craton, Brazil. J Geol Surv Braz 6:217–237

Amaya López C, Restrepo Álvarez JJ, Weber Scharff M, Cuadros FA, Francisquini Botelho M, Ibáñez Mejía M, Maya Sánchez M, Pérez Parra OM, Ramírez Cárdenas C (2020) The Guaviare Complex: new evidence of Mesoproterozoic (ca. 1.3 Ga) crust in the Colombian Amazonian Craton. Boletín Geológico 47:5–34. https://doi.org/10.32685/0120-1425/boletingeo.47.2020.502

Anandbahadoer-Mahabier R, De Roever EWF (2019) The Caicara-Dalbana Belt, a Belt of Felsic and Intermediate Metavolcanics of 1.99 Ga in the Guiana Shield, and Probably Across, in the Guapore Shield. Proceedings SAXI- XI Inter Guiana Geological Conference 2019: Paramaribo. Suriname, Mededeling Geologisch Mijnbouwkundige Dienst Suriname 29:7–13

Barbosa NA, Fuck RA, Souza VS, Dantas EL, Tavares Júnior SS (2021) Evidence of a Palaeoproterozoic SLIP, northern Amazonian Craton, Brazil. J S Am Earth Sci 111:103453

Barron CN (1969) Notes on the stratigraphy of Guyana. Proceedings Seventh Guiana Geological Conference, Paramaribo, 1966. Records Geological Survey Guyana 6, II: 1–28

Bastos Neto AC, Pereira VP, Ronchi LH, de Lima EF, Frantz C (2009) The world-class Sn-Nb-Ta, F (Y, REE, Li) deposit and the massive cryolite associated with albine-enriched facies of the Madeira A-type granite, Pitinga mining district, Amazonas state, Brazil. The Canadian Mineralogist 47:1329-1357

Berrangé J (1977) The Geology of southern Guyana, South America. Institute of Geological Sciences Overseas Memoir 4, 112

Bonilla-Pérez A, Frantz JC, Charão-Marques J, Cramer T, Franco-Victoria JA, Mulocher E, Amaya-Perea Z (2013) Petrografía, Geoquímica y Geocronología del Granito de Parguaza en Colombia. Boletin de Geología 35(2):83–104. https://doi.org/10.18273/revbol

Bonilla-Pérez A, Frantz JC, Charão-Marques J, Cramer T, Franco-Victoria JA, Amaya-Perea Z (2016) Magmatismo Rapakivi En La Cuenca Media Del Río Inírida, Departamento de Guainía, Colombia. Boletin de Geología 38(1):17–32. https://doi.org/10.18273/revbol.v38n1-2016001

Bonilla-Pérez A, Cramer T, de Grave J, Alessio BL, Glorie S, Kroonenberg SB (2021) The NW Amazonian Craton in Guainía and Vaupes departments, Colombia: Transition between orogenic to anorogenic environments during the Paleo-Mesoproterozoic. Precambr Res 360:106223

Bosma W, Kroonenberg SB, Maas K, de Roever EWF (1983) Igneous and metamorphic complexes of the Guiana shield in Suriname. Geol Mijnbouw 62:241–254

Cordani UG, Tassinari CCG, Teixeira W, Basei MAS, Kawashita K (1979) Evolução tectônica da Amazonia com base nos danos geocronológicos. 2ndo Congreso Geológico Chileno 139–148

Cordani UG, Fraga LM, Reis N, Tassinari CCG, Brito-Neves BB (2010) On the origin and tectonic significance of the intra-plate events of Grenvillian-type age in South America: A discussion. J S Amn Earth Sci 29:143–159

Cordani UG, Sato K, Sproessner W, Fernandes FS (2016) U-Pb Zircon ages of rocks from the amazonas territory of Colombia and their bearing on the tectonic history of the NW sector of the Amazonian Craton. Braz J Geol 46(1):5–35

Delor C, de Roever EWF, Lafon JM, Lahondère D, Rossi Ph, Cocherie A, Guerrot C, Potrel A (2003b) The Bakhuis ultra-high temperature granulite belt Suriname : II implications for late Trans-Amazonian crustal stretching in a revised Guiana Shield framework. Géologie de la France 2–4:207–230

De Roever EWF, Lafon J-M, Delor C, Cocherie A, Guerrot C (2015) Orosirian magmatism and metamorphism in Suriname: new geochronological constraints. Contribuições á Geologia Da Amazônia 9:359–372

De Roever EWF, Lafon JM, Delor C, Rossi P, Cocherie A, Guerrot C, Potrel A (2003b) The Bakhuis Ultra-high temperature granulite belt : I Petrological and geochronological evidence for a counterclockwise P-T path at 2.07–2.05 Ga. Géologie de la France 2003, 2–4:175–205

De Roever EWF, Beunk FF, Yi K, de Groot K, Klaver M, Nanne JAM, van de Steeg W, Thijssen ACD, Uunk B, Vos H, Davies GR, Brouwer FM (2019) The Bakhuis Granulite Belt in W Suriname, its development and exhumation. In: Proceedings 11th Inter Guiana Geological Conference, Paramaribo, Mededeling Geologisch Mijnbouwkundige Dienst Suriname 53–58

De Roever EWF, Beunk F, Yi K, Donker R-J, Van de Steeg W, Uunk B, Davies GR, Brouwer FM (2022) Ultrahigh-temperature metamorphism in the Bakhuis Granulite Belt (Suriname). In: Proceedings SAXI-XII Inter-Guiana Geological Conference 2022: Georgetown, Guyana, pp 54–59

Fraga LM, Reis NJ, Dall'Agnol R (2009a). The Cauarane-Coeroene belt, the main tectonic feature of the central Guyana Shield, northern Amazonian Craton. SBG Núcleo Norte, Simpósio de Geologia da Amazônia 11, Manaus, Expanded Abstract, pp 3

Fraga LM, Macambira MJB, Dall'Agnol R, Costa JBS (2009b) 1.94–1.93 Ga charnockitic magmatism from the central part of the Guiana Shield, Roraima, Brazil: single zircon evaporation data and tectonic implications. J S Am Earth Sci 27:247–257

Fraga LM, Cordani U, Kroonenberg S, de Roever E, Nadeau S, V. Maurer VC (2017) Shrimp U-Pb new data on the high grade supracrustal rocks of the Cauarane-Coeroene belt- insights on the tectonic evolution of the Guiana Shield. Anais do 15° Simpósio de Geologia da Amazônia, Belém, pp 486–490

Fraga LM, Cordani, U (2019) Early Orosirian tectonic evolution of the Central Guiana Shield: insights from new U-Pb SHRIMP data. In: Proceedings 11th Inter Guiana Geological Conference, Paramaribo, Mededeling Geologisch Mijnbouwkundige Dienst Suriname, pp 59–62

Fraga LM, Reis NJ, Klein E, Dreher A, Scandolara J (2022). The Orocaima Igneous Belt and the 1.99–1.96 Ga SLIP in the Amazonian Craton. Proceedings 12th Inter Guiana Geological Conference, Georgetown, Guyana, pp 60–65

Fraga LM, Cordani UG, Dreher AM, Sato K, Reis NJ, Nadeau S, De Roever E, Kroonenberg S, Camara Maurer VC (2024) Early Orosirian belts of the central Guiana Shield, northern Amazonian Craton: U-Pb geochronology and tectonic implications Precambrian Research, pp 407

Galvis J, Huguett A, Ruge P (1979) Geología de la Amazonia Colombiana. Boletín Geológico INGEOMINAS 22(3):3–86

Gibbs AK, Barron CN (1983) The Guiana Shield reviewed. Episodes 7–14

Gibbs AK, Barron CN (1993) Geology of the Guiana shield. Oxford University Press, pp 246

Hackley PC, Urbani F, Karlsen AW, Garrity, CP (2005) Geologic shaded relief map of Venezuela. USGS Open File Report 2005–1038

Heinonen AP, Fraga LM, Rämö OT, Dall'Agnol R, Mänttäri I, Andersem T (2012) Petrogenesis of the igneous Mucajaí AMG complex, northern Amazonian craton. Geochemical, U-Pb geochronological, and Nd–Hf–O isotopic constraints. Lithos 151:17–34

Huguett A, Galvis J, Ruge P (1979) Geología. In: La Amazonia colombiana y sus recursos. Proyecto Radargramétrico del Amazonas, Bogotá 29–92

Ibáñez-Mejía M, Ruiz J, Valencia VA, Cardona A, Gehrels GE, Mora AR (2011) The Putumayo Orogen of Amazonia and its implications for Rodinia reconstructions: New U-Pb geochronological insights into the Proterozoic tectonic evolution of Northwestern South America. Precambr Res 191:58–77

Ibáñez-Mejía M, Cordani UG (2020) U–Pb Geochronology and Hf–Nd–O isotope geochemistry of the Paleo– to Mesoproterozoic basement in the westernmost Guiana Shield. In: Gómez, J. and Mateus–Zabala, D. (eds), The Geology of Colombia, Volume 1 Proterozoic –Paleozoic. Servicio Geológico Colombiano, Publicaciones Geológicas Especiales 35:65–90 https://doi.org/10.32685/pub.esp.35.2019.04

Klaver M, De Roever EWF, Nanne JAM, Mason PRD, Davies GR (2015) Charnockites and UHT metamorphism in the Bakhuis Granulite Belt, western Suriname: Evidence for two separate UHT events. Precambr Res 262:1–19

Klaver M, de Roever EWF, Thijssen ACD, Bleeker W, Söderlund U, Chamberlain K, Ernst R, Berndt J, Zeh A (2016) Mafic magmatism in the Bakhuis Granulite Belt (western Suriname): relationship with charnockite magmatism and UHT metamorphism. GFF 138(1):203–218. https://doi.org/10.1080/11035897.2015.1061591

Klein EL, Almeida ME, Rosa-Costa LT (2012) The 1.89–1.87 Ga Uatumã Silicic Large Igneous Province, northern South America. In: Large Igneous Provinces Commission (http://www.lar geigneousprovinces.org) 1–14

Kroonenberg SB (1975) Geology of the Sisa Creek area, SW Suriname. Geol Mijnb Dienst Sur Med 23:102–125

Kroonenberg SB (1976) Amphibolite-facies and granulite-facies metamorphism in the Coeroeni-Lucie área, SW Surinam. Thesis Amsterdam, Mededelingen Geologisch Mijnbouwkundige Dienst Suriname 25:109–289

Kroonenberg SB (1982) A Grenvillian granulite belt in the Colombian Andes and its relation to the Guiana Shield. Geologie en Mijnbouw 61: 325–333

Kroonenberg SB (2019b) The Proterozoic Basement of the Western Guiana Shield and the Northern Andes. In: Cediel F, Shaw RP (eds.), Geology and Tectonics of Northwestern South America, Frontiers in Earth Sciences, Springer, pp 115–192

Kroonenberg SB, de Roever EWF (2010) Geological Evolution of the Amazonian Craton, in: Amazonia, Landscape and Species Evolution. In: Hoorn C, Wesselingh F.P (eds), Wiley, pp 9–28

Kroonenberg SB, de Roever EWF, Fraga LM, Reis NJ, Faraco MT, Cordani UG, Lafon J-M, Wong TE (2016) Paleoproterozoic evolution of the Guiana Shield in Suriname–a revised model. Neth J Geosci-Geol En Mijnb 95:491–522

Kroonenberg S, Mason PRD, Kriegsman LM, de Roever EWF, Wong TE (2019) Geology and mineral deposits of the Guiana Shield. In: Proceedings 11th Inter Guiana Geological Conference, Paramaribo. Mededeling Geologisch Mijnbouwkundige Dienst Suriname 29:111–115

Lafon JM, Rosa-Costa L, Milhomem Neto JM (2019) Sr-Nd-Hf isotopic tracing of Archean continental crust in the Brazilian part of the Southeastern Guyana Shield: A review. In: Proceedings 11th Interguiana Geological Conference 2019: Paramaribo, Suriname. Mededeling Geologisch Mijnbouwkundige Dienst Suriname 29:121–124

Loemban Tobing DP (1969). Geology of the Avanavero area in Western Surinam. In: Proceedings of the 7th Guiana Geological Conference, Paramaribo, 1966. Verhandelingen van het Koninklijk Nederlands Geologisch Mijnbouwkundig Genootschap 27:33–48

Martín-Bellizzia C (1972) Paleotectónica del Escudo de Guayana. Memoria de la 9na conferencie geológica Inter-Guayanas. Boletín de Geología (Caracas). Publicación Especial No. 6:251–304

Montalvão RMG (1975) Grupo Uatumã no Craton Guianês. Anais Décima Conferência Geológica Interguianas (Belém), 286–339

Moyano Nieto IE, Prieto GA (2021) Structural signatures of the Amazonian Craton in eastern Colombia from gravity and magnetometry data interpretation. Tectonophysics 800: 228705

Nanne JAM, de Roever EWF, de Groot K, Davies GR, Brouwer FM (2020) Regional UHT metamorphism with widespread, primary CO_2-rich cordierite in the Bakhuis Granulite Belt, Surinam: A feldspar thermometry study. Precambr Res 350:105894

Priem HNA, Hebeda EH, Boelrijk NAIM, Verschure RH (1968) Isotope age determinations on Surinam rocks, 3. Proterozoic and Permo-Triassic basalt magmatism in the Guiana Shield. Geol Mijnbouw 47:17–20

Priem HNA, Boelrijk NAIM, Hebeda EH, Verdurmen EAT, Verschure RH (1971) Isotopic ages of the Trans-Amazonian acidic magmatism and the Nickerie Episode in the Precambrian basement of Surinam, South America. Geol Soc Am Bull 82:1667–1680

Priem HNA, Boelrijk NAIM, Hebeda EH, Kroonenberg SB, Verdurmen EAT, Verschure RH (1977) Isotopic ages in the high grade metamorphic Coeroeni Group, southwestern Suriname. Geol Mijnbouw 56(155):160

Priem HNA, Andriessen PAM, Boelrijk NAIM, de Boorder H, Hebeda EH, Huguett A (1982) Geochronology of the Precambrian in the Amazonas region of southwestern Colombia, western Guiana shield. Geol Mijnbouw 61:229–242

Ramgrab GE, Santos JOS (1974) O grupo Uatumã. Anais do 28° Congresso Brasileiro de Geologia: 87–94.

Reis NJ, de Faria MSG, Fraga LM, Haddad RC (2000) Orosirian calc-alkaline volcanism and the Orocaima event in the Northern Amazônian Craton, Eastern Roraima State. Brazil. Revista Brasileira De Geociências 30(3):38–383

Reis NJ, Teixeira W, D'Agrella-Filho MS, Bettencourt JS, Ernst RE, Goulart LEA (2021) Large igneous provinces of the Amazonian Craton and their metallogenic potential in Proterozoic times. In: Srivastava R.K, Ernst R.E, Buchan K.L, de Kock M (eds) Large Igneous Provinces and their plumbing systems. Geological Society, London, Special Publications 518:493–529

Santos JOS (2003) Geotectônica dos Escudos das Guianas e Brasil-Central. Geotectonics of the Guyana and Central Brazilian Shields. In: Bizzi L.A, Schobbenhaus C, Vidotti R.M, Gonçalves E.J.H (eds.) Geologia, Tectônica e Recursos Minerais do Brasil 169–226

Santos JOS, Hartmann LA, Gaudette HE, Groves DI, McNaughton NJ, Fletcher IR (2000) A new understanding of the provinces of the Amazon Craton based on integration of field mapping and U-Pb and Sm-Nd geochronology. Gondwana Res 3:453–488

Santos JOS, Hartmann LA, Faria MS, Riker SR, Souza MM, Almeida ME, McNaughton NJ, (2006a) Compartimentação do Cráton Amazonas em províncias: avanços ocorridos no período 2000–2006. Simpósio de Geologia da Amazônia, vol. 9, Sociedade Brasileira de Geologia, Belém, Brazil, Resumos Expandidos, CD ROM

Santos JOS, Faria MS, Riker SR , Souza MM, Hartmann LA, Almeida ME , McNaughton NJ, Fletcher IR (2006b) A faixa colisional K'mudku (idade Grenvilliana) no norte do Cráton Amazonas: reflexo intracontinental do Orógeno Sunsás na margem ocidental do cráton. Simpósio de Geologia da Amazônia, vol. 9, Sociedade Brasileira de Geologia, Belém, Brazil, Resumos Expandidos, CD ROM

Sidder GB, Mendoza V (1995) Geology of the Venezuelan Guayana Shield and its relation to the geology of the entire Guayana Shield. In: Sidder et al (eds) Geology and Mineral Deposits of the Venezuelan Guiana Shield. U.S. Geological Survey Bulletin 2124, B1-B41

Tassinari CCG, Macambira MJB (1999) Geochronological provinces of the Amazonian Craton. Episodes 22(3):174–182

Teixeira W, Tassinari CCG, Cordani UG, Kawashita K (1989) A review of the geochronology of the Amazonian Craton: tectonic implications. Precambr Res 42:213–227

Verhofstad J (1971) The geology of the Wilhelmina Mountains in Suriname, with special referene to the occurrence of Precambrian Ash-flow tuffs. Thesis Univ. Amsterdam. Mededeling Geologisch Mijnbouwkundige Dienst Suriname 21:9–97

Yakymchuk C, Kirkland CL, Clark C (2018) Th/U ratios in metamorphic zircon. J. Metamorphic Geology 36:715–737

Chapter 9
Synthesis: The Trans-Amazonian Orogeny

Abstract This chapter offers a review of the main stages of the Trans-Amazonian Orogeny that shaped a large part of the Guiana Shield, based on recent geochronological data and geotectonic interpretations, and stimulated by renewed international cooperation.

From 2019 onward geological research in the Guianas has received an enormous stimulus through the South American Exploration Programme SAXI (saxiproject.org). The Australian non-profit organisation AMIRA has contracted a consortium of several major mining companies to sponsor fundamental scientific research in the Guianas from 2019 to 2024. The SAXI programme has resulted in numerous research projects in local and international universities and geological surveys, especially focused on gold exploration. Many results are still awaiting publication, but a number of extended abstracts have been published in the proceedings of the SAXI-sponsored 11th and 12th Inter Guiana Geological Conferences in Paramaribo in 2019 and Georgetown in 2022, respectively. The project has inspired me to put together the following synthesis of the Trans-Amazonian Orogeny. The Trans-Amazonian Orogeny between 2.26 and 1.95 Ga is the most pervasive event that shaped the main blocks of the Guiana Shield (Fig. 9.1). Several stages can be discerned in its development.

9.1 The Maronian Greenstone Belt

(1) Oceanic stage (2.26–2.18 Ga?). Spreading ridge magmatism, extrusion of pillowed tholeiitic basalts and locally magnesian volcanics with komatiitic geochemistry (Gibbs 1980; Vanderhaeghe et al. 1998; Capdevila et al. 1999; Naipal et al. 2019, 2022a), though spinifex textures have never been recorded. Velázquez et al. (2011) and Naipal et al. (2023) both argue for an oceanic plateau origin on the basis of geochemical arguments; this makes an open ocean setting more plausible than a back-arc setting. Reliable age data for these series are not

© The Author(s), under exclusive license to Springer Nature Switzerland AG 2025
S. Kroonenberg, *The Changing Framework of the Guiana Shield*, SpringerBriefs in Earth System Sciences, https://doi.org/10.1007/978-3-031-86334-9_9

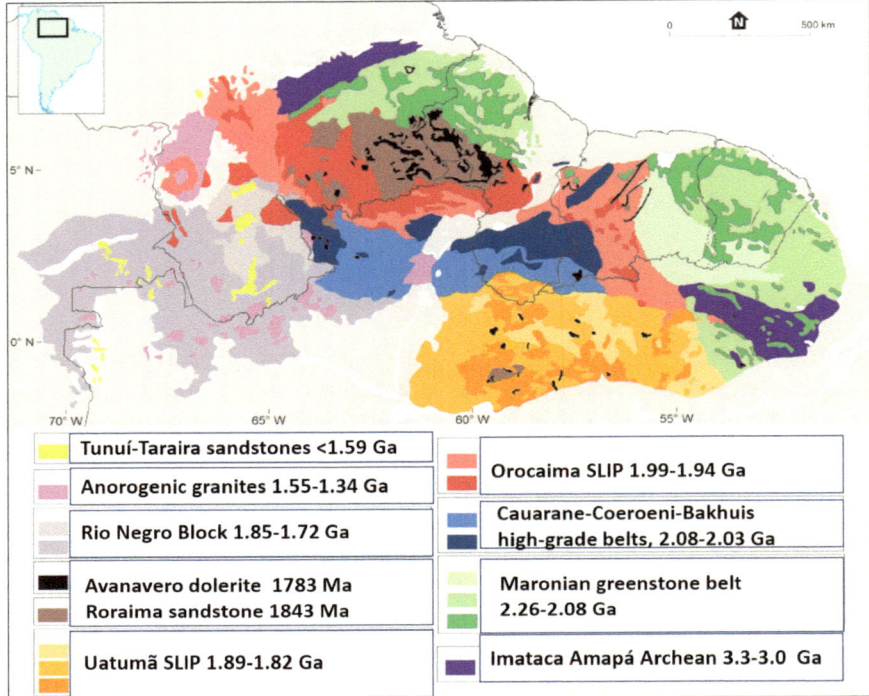

Fig. 9.1 The main building blocks of the Guiana Shield: as Fig. 1.3, but now with a simpler colour scheme (modified after Kroonenberg et al. 2016)

available, possibly apart from the 2.26 Sm-Nd isotope age for actinolite schists and amphibolites from Amapá (McReath and Faraco 1985). Vanderhaeghe et al. (1998) and Delor et al. (2003a) cite an age of 2216 ± 4 Ma for a Fe-gabbro near Cayenne as the earliest rock-forming event in the Trans-Amazonian orogeny.

(2) Southwards subduction and island arc magmatism, including calcalkaline andesitic to rhyolitic volcanism and volcaniclastic sedimentation, diapiric TTG and (ultra)mafic plutonism (2.18–2.12 Ga). Delor et al. (2003a) give 2.14–2.15 Ga ages from metarhyolites in French Guiana, Daoust (2016) obtained a U-Pb zircon age of 2.16 Ga for the Mayo metarhyolite in the Rosebel area of Suriname, and an andesite dyke in Guyana gave 2.15 Ga (Tedeschi et al. 2020). Coeval chemical seafloor sedimentation gave rise to cherts, manganese deposits and itabirites.

TTG ages between 2.18 and 2.12 Ga have been obtained in Suriname (Daoust 2016; Ramlal et al. 2019), between 2.18 and 2.13 Ga in French Guiana (Delor et al. 2003a; Combes et al. 2022). Comparable intrusives in Guyana are somewhat younger, between 2.12 and 2.08 Ga (Tedeschi et al. 2020; Thébaud and Tedeschi 2022).

(Ultra)mafic intrusions into the tholeiitic and ultramafic volcanics have an arc-like geochemistry (Veenstra 1983; Naipal et al. 2019). Delor et al. (2003a) published a 2.15 Ga age for the Tampok gabbro in French Guiana, Naipal (pers. comm) obtained an baddeleyite/zircon in situ U-Pb age of 2.13 Ga for the Piqué gabbro in eastern Suriname, and Borba de Carvalho et al. (2022) reported four zircon and baddeleyite ages between 2.10 and 2.24 from cumulate mafic rocks from Matthews Ridge, Guyana. Apparently there is no specific sequence in the appearance of mafic, intermediate or felsic magmas.

(3) Submarine turbiditic sedimentation: rhythmic deposition of graded immature greywackes, mudstones and some conglomerates, probably triggered by shelf instability along the island arcs. Maximum grain size and metamorphic grade seems to diminish from Suriname to Guyana. Tonalite clasts testify of previous exhumation of TTG bodies (Naipal and Kroonenberg 2016). Detrital ages span a continuum from 2.16 to 2.73 Ga in the Coppename River (Wijngaarde et al. 2019) and present age clusters around 2.12, 2.60 and 3.02 Ga in the Rosebel mine (Daoust 2016), both in Suriname.

(4) Epicontinental fluvial sedimentation: Polymict conglomerates and immature arenites to mature quartzarenites show crossbedding, testifying of fluvial deposition. In French Guiana they occupy narrow pull-apart basins at the borders of the Orapu basin, where they unconformably overlie the turbidites (Vanderhaeghe et al. 1998; Delor et al. 2003a). Vanderhaeghe et al. (1998) interpret them as having been deposited by braided rivers and debris flows. In Suriname they form the centre of the main greenstone belt synclinorium, but contacts with older formations have not been observed. Daoust (2016) gives detrital zircon ages from several samples with peaks around 2.09, 2.11, ~2.15 and ~2.60 Ga, and Naipal et al. (2022b) gives peaks 2.11, 2.65 and 3.1 Ga from a borehole near the Rosebel mine. An intercalated rhyolitic tuff layer gave 2.12 Ga (Ramlal et al. 2019). This stage is absent in Guyana, Venezuela and Amapá.

(5) Late potassic S-type magmatism. In northeastern Suriname an northern French Guiana a series of S-type metaluminous to peraluminous granites intruded, often with abundant muscovite, locally with garnet and with numerous metasedimentary enclaves. They produce conspicuous aureoles of coarse staurolite-garnet schists, considered as contactmetamorphosed turbidite sequences. The schists in turn constitute the wall-rock of swarms of pegmatites, partly of the LCT (lithium, cesium, tantalum) type, proceeding from the S-type granite plutons. In French Guiana they have been dated between 2.08 and 2.06 Ga (Delor et al. 2003a), and in Suriname between 2.11 and 2.07 Ga (Kromopawiro et al. 2019), while also in the Cayenne area 2.09–2.08 Ga granites are reported near the contact with the Orapu/Regina sedimentary basins (Vanderhaeghe et al. 1998).

(6) The ~2.08 Ga migmatitic transition zones from the low-grade metavolcanic-metasedimentary parts of the greenstone belt to 2.09 Ga diatexitic granitoids in Suriname (Sastrohardjo et al. 2022) resemble the Laussat Complex in French Guiana (2.17 Ga, Delor et al. 2003a; Choubert's 1974 Granite Galibi), the Bartica

Gneiss in Guyana (2.11 Ga, Tedeschi et al. 2020) and the Supamo Complex in Venezuela at 2.13–2.11 Ga (Hildebrand et al. 2014).

Denudational intervals within the sequences are evident from the presence of volcanic and tonalitic clasts in the turbidites, and of phyllite and granitoid clasts in the fluvial sediments. Each of these intervals represents a period of deformation, metamorphism and denudation, as postulated by Delor et al. (2003a,b). Metamorphism is usually in the greenschist facies, except around TTG bodies and S-type granitoids, where they present amphibolite-facies assemblages.

There are some striking differences between the western limb of the Maronian greenstone belt in Venezuela, Guyana and northwestern Suriname on one hand, and the eastern limb in northeastern Suriname, French Guiana and Amapá. Greywackes in the western limb show lower greenschist-facies metamorphism than in the eastern limb (Tedeschi et al. 2022). Epicontinental sediments and late S-type granites with their pegmatites are absent from the western limb, and the eastern TTG bodies are generally older than the western ones. Bardoux (2022) regards them as different basins, the Cuyuni and Marowijne megabasins, on the basis of differences between their gold deposits. Copper in the gold-bearing districts of the western limb is disseminated, but of VMS type in those of the eastern limb (Kroonenberg 2023). Also Beunk et al. (2021) consider them as two greenstone belts on both sides of the Bakhuis Granulite Belt, conforming an orogenic syntax.

9.2 The High-Grade Belts

In fact there a two types of Trans-Amazonian high-grade belts in the Guiana Shield: those with an Archean core, i.e. the Imataca belt in Venezuela and the Amapá belt in northeastern Brazil, and the Rhyacian belts without Archean ancestry, the Cauarane-Coeroeni-Bakhuis belts in the central part of the shield. Both types underwent granulite-facies metamorphism in the same time interval in the Trans-Amazonian Orogeny: between 2.05 and 1.98 Ga in the Imataca belt (Tassinari et al. 2004), between 2.10 and 2.04 Ga in the Amapá belt (Rosa-Costa et al. 2006, 2008; Milhomem Neto et al. 2022), and between 2.09 and 2.03 Ga in the Cauarane-Coeroeni-Bakhuis belt. The 2.09–2.03 ages of the UHT granulite-facies metamorphism in the Bakhuis belt have been summarized by De Roever et al. (2019). I consider the allegedly detrital zircon ages between 2.09 and 2.03 Ga in the Cauarane-Coeroeni belt of Fraga et al. (2024) as indicating the age of the first phase of high-grade metamorphism, hence coeval with metamorphism in the Bakhuis belt, as argued in chapter 8.1.

For the origin of the Bakhuis Belt Delor et al. (2003b) design a Late Rhyacian phase of crustal stretching, leading initially to pull-apart rift formation and deposition of the supracrustals of the Bakhuis belt and finally high-grade metamorphism with an anti-clockwise cooling path (De Roever et al. 2003), which pleads against a collisional origin (Delor et al. 2003a; Kroonenberg et al. 2016). The equally supracrustal

Cauarane-Coeroeni Belt displays a similar anticlockwise cooling pattern (Kroonen-berg 1976; Kroonenberg et al. 2016), and therefore an origin by a collision between an Orosirian magmatic arc and a previously existing Rhyacian plate as envisaged by Fraga et al. (2024) seems unlikely. Whether the triple-junction pattern suggested by the configuration of the three high-grade belts can also be explained by Delor's pull-apart mechanism is still an open question.

In the Imataca belt no anti-clockwise cooling path has been recorded, all granulite-facies cordierite-garnet-sillimanite bearing assemblages appear in equilibrium with each other (Dougan 1974). Cordierite gneisses have been reported from the Neoarchean Cupixi domain in Amapá (Rosa-Costa et al. 2003) and interpreted as having been metamorphosed during a collisional stage of the Trans-Amazonian Orogeny (Rosa-Costa et al. 2009). Any geodynamic reconstruction of this phase of the Trans-Amazonian Orogeny should therefore consider a mechanism for the simultaneously occurring granulite-facies metamorphism in both the Archean and the Rhyacian belts.

The relation of the high-grade belts with the Maronian greenstone belt is not always straightforward. In Amapá the Rhyacian Carecuru granite-greenstone belt was a magmatic arc accreted to the southwestern border of the Archean Amapá Block during the Trans-Amazonian event (Rosa Costa et al. 2006, 2009). Hildebrand et al. (2014) consider the greenstone belt in Venezuela as allochthonous, thrusted on top of the Imataca belt along the continental Guri fault zone. In northern Suriname aeromagnetic imagery suggests that, contrary to the statement by Beunk et al. (2021), the younger Bakhuis belt is truncated in the north by the older greenstone belt. This was already remarked by Gibbs and Barron (1993, their Fig. 2.5, p. 38), and confirmed later by Girjasing et al. (2019). Possibly the continental Northern Suriname Shear Zone postulated by Voicu et al. (2001) played a role in this anomalous situation.

9.3 The Orocaima Belt

The period between, broadly speaking, 2.0 and 1.95 Ga saw an exceptional series of events. First of all, the generation of an enormous volume of felsic magma covering a surface area of over 200,000 km^2, i.e. several millions of km^3 intruded and extruded rock. With right this can be called a Siliceous Large Igneous province (Reis et al. 2021; Barbosa et al. 2021). It extends over a distance of 2000 km from the Colombian-Venezuelan border to deep into southeastern Brazil, closely following the southern border of the Maronian greenstone belt. This on itself already suggests a relation with the Trans-Amazonian Orogeny. Extrusion of ignimbritic rhyolites usually precedes the intrusion of granitoid bodies, as the ignimbrites show recrystallisation and contact metamorphism around granitoid plutons (Verhofstad 1971; Bosma et al. 1983; De Roever et al., 2015) The 1.98 Ston quartzites and conglomerates which underlie the ignimbrites in northwestern Suriname (Chapter 8.2) also show contact meta-morphism around a granite intrusion (Loemban Tobing 1969). A possible caldera, maybe one of many, has been identified in the northern part of Roraima province, the Tepequém caldera filled with Roraima sediments and surrounded by concentric subvolcanic granites (Reis et al. 2003, 2009; Fraga et al. 2010; Barbosa et al. 2021).

What triggered the generation of such a large volume of felsic melt? Fraga et al. (2024) consider it as a late- to postmagmatic belt related to a continental collision after northwards subduction in the Cauarane-Coeroeni belt, and Kroonenberg et al. (2016) suggest it to be a third post-collisional magmatic phase of the Trans-Amazonian Orogeny. Anandbahadoer-Mahabier and De Roever (2019) and Barbosa et al. (2021) also favour a subduction-related origin on geochemical grounds. The latter authors remark: 'However, this signature is understood as a component inherited from the recycling of rocks generated at the end of the Transamazonian Orogeny. The proposed context for emplacing Orosirian volcanism in the Guiana Shield is post-collisional with the contribution of a subduction component or the reflection of an extension configuration located in a general convergent context.'

That is an interesting suggestion. Ibáñez-Mejía et al. (2022) identified a radial pattern of 1.99–1.96 Ga tholeiitic dyke swarms in northern Venezuela and northern Suriname, geographically and geochronologically related to the Orocaima SLIP, suggesting a plume-related origin for these mafic magmas. Heat transfer from an impacting mafic plume triggered large-scale melting of the Amazonian lithosphere resulting in the SLIP development. The small (ultra)mafic 1.98 Ga plutons such as the Appinitic Suite in Guyana and the Lucie gabbro in Suriname that punctured through the felsic volcanic covers could well be related to this mafic plume. The 'general convergent context' of Barbosa et al. (2021) would then explain the distribution of the SLIP parallel to the Maronian greenstone belt, and possibly also some of the metamorphic effects described in par. 8.2.

9.4 A Synthesis

An interesting study was presented by Tedeschi et al. (2022) in the framework of the SAXI project, in which he compared the stratigraphy of different parts of the greenstone belt in Guyana and Suriname (Fig. 9.2).

It shows the overall similarity in successions, as well as its overall similarity to the stratigraphy in Archean greenstone belts. Thébaud and Tedeschi (2022) showed that TTG U-Pb ages of granitoid intrusions in Guyana correspond well with the deformational episodes obtained by Delor (2003a) for the main Trans-Amazonian magmatic pulses in French Guiana (Fig. 9.3).

Figure 9.4. summarizes the main succession of events in the Trans-Amazonian Orogeny on the basis of the data available up to now. After the Orocaima event the cratonization of the larger part of the Guiana Shield was completed. Denudation followed and led to the deposition of 3000 m of subhorizontal Roraima molassic sediments, interspersed with thin 1.83 Ga ash falls from the developing Uatumã plume along its southern border. The intrusion of the 1.78 Ga Avanavero LIP dolerites into a large part of the shield, but especially in the Roraima basin, marked the largest post-Trans-Amazonian upsurge of mafic magmatism. At the same time the Querarí

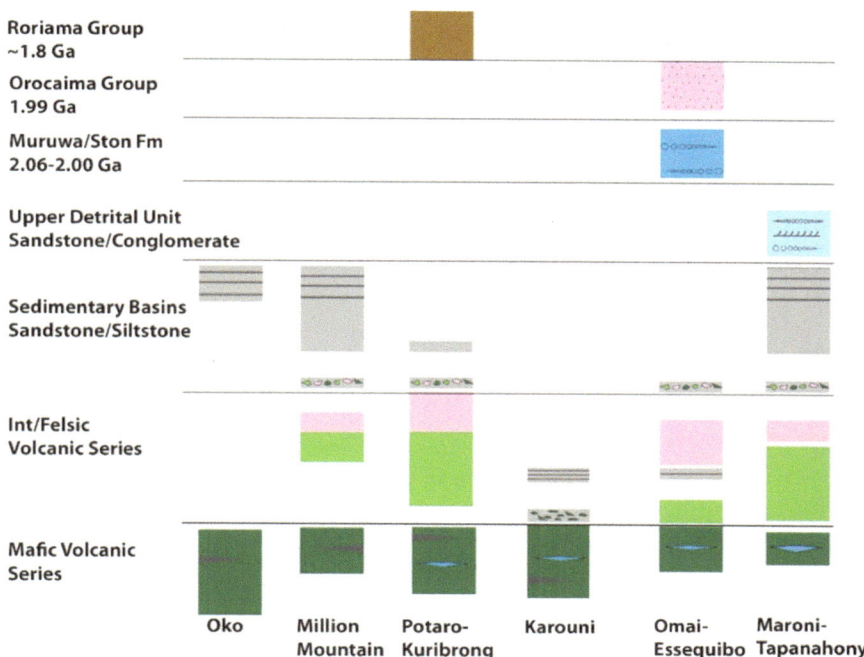

Fig. 9.2 Lithostratigraphy of the greenstone belt supracrustals in different sectors in Guyana and Suriname (Tedeschi et al. 2022)

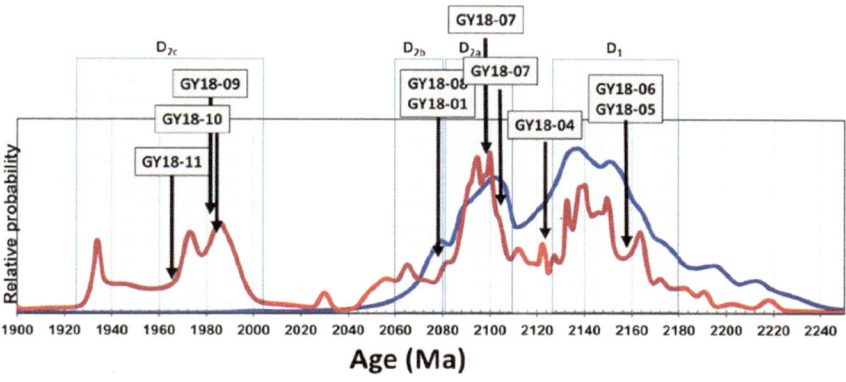

Figure 2: Probability diagram of a compilation of U-Pb zircon geochronology from the Guiana Shield (Tedeschi et al., 2020) with the interpreted orogenic stages (Delor et al., 2003) and results of this study.

Fig. 9.3 A comparison of TTG U-Pb zircon ages of granitoid intrusives in Guyana (GY numbers, red curve) with the stages of the Trans-Amazonian Orogeny of Delor et al. (2003b, blue curve). (Thébaud and Tedeschi 2022)

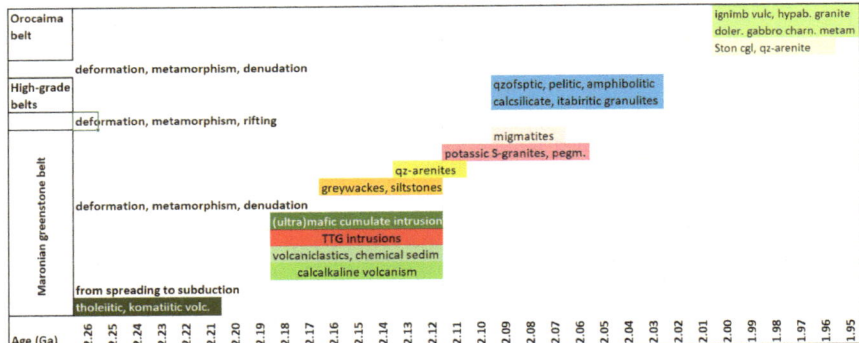

Fig. 9.4 A simplified time schedule of the Trans-Amazonian Orogeny in the Guiana Shield

Orogeny along the western flank of the shield developed, leading to the accretion of the Río Negro Block, followed by anorogenic rapakivi and other granitoid plutonism, and again molassic sandstone deposition in the folded younger Tunuí-Taraira sandstones. The collision of Laurentia with Amazonia around 1.2–1.0 Ga during the Grenvillian or Putumayo Orogeny led to widespread mylonitization, thermal resetting of mica ages and localised intrusion of alkaline and carbonatite plugs in the shield. The rest is history.

References

Anandbahadoer-Mahabier R, De Roever EWF (2019) The Caicara-Dalbana Belt, a Belt of Felsic and Intermediate Metavolcanics of 1.99 Ga in the Guiana Shield, and Probably Across, in the Guapore Shield. In: Proceedings SAXI-XI Inter Guiana Geological Conference 2019: Paramaribo, Suriname, Mededeling Geologisch Mijnbouwkundige Dienst Suriname 29:7–13

Barbosa NA, Fuck RA, Souza VS, Dantas EL, Tavares Júnior SS (2021) Evidence of a Palaeoproterozoic SLIP, northern Amazonian Craton, Brazil. J S Am Earth Sci 111:103453

Bardoux M (2022) OCSH, IH and IR Rhyacian gold deposits of the Guiana Shield SAXI-XII Inter-Guiana Geological Conference 2022: Georgetown, Guyana 17

Beunk FF, de Roever EWF, Yi K, Brouwer FM (2021) Structural and tectonothermal evolution of the ultrahigh-temperature Bakhuis Granulite Belt, Guiana Shield, Surinam: Palaeoproterozoic to recent. Geosci Front 12:677–692

Borba de Carvalho B, Cousens B, Hunter C, Chamberlain, K, Ernst RE (2022) Paleoproterozoic U-Pb ages and geochemistry of mafic and ultramafic rocks of Matthews Ridge, Guyana: A comparison with other Paleoproterozoic occurrences within the Guiana Shield. In: Proceedings SAXI- XII Inter-Guiana Geological Conference, Georgetown, Guyana, pp 40–44

Bosma W, Kroonenberg SB, Maas K, de Roever EWF (1983) Igneous and metamorphic complexes of the Guiana shield in Suriname. Geol Mijnbouw 62:241–254

Capdevila R, Arndt N, Letendre J, Sauvage J-F (1999) Diamonds in volcaniclastic komatiite from French Guiana. Nature 399:56–458. https://doi.org/10.1038/20911

Combes V, Eglinger A, André-Mayer A-S, Teitler Y, Jessell M, Zeh A, Reisberg L, Heuret A, Gibert
    P (2022) Integrated geological-geophysical investigation of gold-hosting Rhyacian intrusions
    (Yaou, French Guiana), from deposit-to district-scale. J S Am Earth Sci 114:103708
Daoust C (2016), Caractérisation stratigraphique, structurale et géochimique du District minéralisé
    de Rosebel (Suriname) dans le Cadre de l'évolution géodynamique du Bouclier Guyanais, PhD
    thesis, Université du Québec á Montréal (Montréal): 330
Delor C, Lahondère D, Egal E, Lafon JM, Cocherie A, Guerrot C, de Avelar V (2003a) Transamazo-
    nian crustal growth and reworking as revealed by the 1:500,000-scale geological map of French
    Guiana. Géologie de la France 2–4:5–57
Delor C, de Roever EWF, Lafon JM, Lahondère D, Rossi Ph, Cocherie A, Guerrot C, Potrel A
    (2003b) The Bakhuis ultra-high temperature granulite belt Suriname : II implications for late
    Trans-Amazonian crustal stretching in a revised Guiana Shield framework. Géologie de la France
    2–4:207–230
De Roever EWF, Lafon J-M, Delor C, Cocherie A, Guerrot C (2015) Orosirian magmatism and
    metamorphism in Suriname: new geochronological constraints. Contribuições á Geologia Da
    Amazônia 9:359–372
De Roever EWF, Lafon JM, Delor C, Rossi P, Cocherie A,.Guerrot C, Potrel A (2003) The Bakhuis
    Ultra-high temperature granulite belt : I Petrological and geochronological evidence for a
    counterclockwise P-T path at 2.07–2.05 Ga. Géologie de la France 2–4:175–205
De Roever, EWF, Beunk FF, Yi K, de Groot K, Klaver M, Nanne JAM, van de Steeg W, Thijssen
    ACD, Uunk B, Vos H, Davies GR, Brouwer FM (2019) The Bakhuis Granulite Belt in W
    Suriname, its development and exhumation. In: Proceedings 11th Inter Guiana Geological
    Conference, Paramaribo, Mededeling Geologisch Mijnbouwkundige Dienst Suriname, pp 53–58
Dougan TW (1974) Cordierite gneisses and associated lithologies of the Guri Area, Northwest
    Guayana Shield, Venezuela Contributions to Mineralogy and Petrology 46:169–188
Fraga LMB, Dreher AM, Grazziotin H, Reis NJ, de Farias MSG, Ragatky D (2010) Geologia
    e recursos minerais da folha Vila de Tepequém-NA.20-X-A-lll, Estado de Roraima, escala
    1:100.000. CPRM Brazil, pp 182
Fraga LM, Cordani UG, Dreher AM, Sato K, Reis NJ, Nadeau S, De Roever E, Kroonenberg
    S, VCamara Maurer VC (2024) Early Orosirian belts of the central Guiana Shield, northern
    Amazonian Craton: U-Pb geochronology and tectonic implications Precambrian Research
    407:107362
Gibbs AK (1980) Geology of the Barama-Mazaruni Supergroup of Guyana. Ph.D. thesis, Harvard
    University, Cambridge, Mass., USA, 385
Gibbs AK, Barron CN (1993) Geology of the Guiana shield. Oxford University Press, 246
Girjasing R, Kroonenberg SB, Ramdjal R, Wong TE (2019) Timing of the origin and uplift
    of the Bakhuis-Tambaredjo Horst, Suriname. In: Proceedings 11th Inter Guiana Geological
    Conference, Paramaribo. Mededeling Geologisch Mijnbouwkundige Dienst Suriname 29:67–70
Hildebrand RS, Buchwaldt R, Bowring SA (2014) On the allochthonous nature of auriferous
    greenstones, Guayana shield, Venezuela. Gondwana Res 26:1129–1140
Ibáñez-Mejía M, Ernst R, Söderlund U, Urbani F, Antonio P, Kroonenberg S, Pepper M (2022).
    A newly recognized 1.98 Ga large igneous province (LIP) in the Amazonian Craton and its
    relationship with the coeval Orocaima silicic LIP. Proceedings 12th Inter Guiana Geological
    Conference, Georgetown, Guyana, 76
Kromopawiro S, Kroonenberg SB, Kriegsman LM, Mason PRD (2019) 2.12–2.08 Ga Late-to post-
    collisional peraluminous granitoid magmatism in the Marowijne Greenstone Belt of Suriname.
    In: Proceedings 11th Inter Guiana Geological Conference, Paramaribo. Mededeling Geologisch
    Mijnbouwkundige Dienst Suriname 29:105–109
Kroonenberg SB (1976) Amphibolite-facies and granulite-facies metamorphism in the Coeroeni-
    Lucie área, SW Surinam. Thesis Amsterdam, Mededelingen Geologisch Mijnbouwkundige
    Dienst Suriname 25:109–289
Kroonenberg SB (2023) Copper in the Guiana Shield. Society of Economic Geologists, Conference
    Resourcing the Green Transition, London. Abstract A83

Kroonenberg SB, de Roever EWF, Fraga LM, Reis NJ, Faraco MT, Cordani UG, Lafon J-M, Wong TE (2016) Paleoproterozoic evolution of the Guiana Shield in Suriname-a revised model. Neth J Geosci-Geol En Mijnb 95:491–522

Loemban Tobing DP (1969) Geology of the Avanavero area in Western Surinam. In: Proceedings of the 7th Guiana Geological Conference, Paramaribo, 1966. Verhandelingen van het Koninklijk Nederlands Geologisch Mijnbouwkundig Genootschap 27:33–48

Milhomem Neto JM, Lafon JM, de P. Amaral Ferreira D, Silva Miranda S, Dantas EL (2022) High-grade metamorphism in the central region of Amapá, Northern Brazil: age constraints from in situ U-Pb dating of monazite and zircon. SAXI- XII Inter-Guiana Geological Conference 2022: Georgetown, Guyana, 105–109

Naipal R, Kroonenberg SB (2016) Provenance signals in metaturbidites of the Paleoproterozoic greenstone belt of the Guiana Shield in Suriname. Neth J Geosci-Geol En Mijnb 95:467–489

Naipal R, Kroonenberg S, Mason PRD (2019) Ultramafic rocks of the Paleoproterozoic greenstone belt in the Guiana Shield of Suriname, and their mineral potential. In: Proceedings 11th Inter Guiana Geological Conference, Mededeling Geologisch Mijnbouwkundige Dienst Suriname 29:143–146

Naipal R, Kroonenberg S, Mason PRD, Kriegsman LM (2022a) The Bemau Ultramafic Complex and the Borgia Hill Chromite Complex: Two contrasting ultramafic complexes in the Paleoproterozoic basement of Suriname. In: Proceedings SAXI- XII Inter-Guiana Geological Conference, Georgetown, Guyana, 116

Naipal R, Kroonenberg S, Kriegsman L, van Bergen M, Mason P (2022b) Hydrothermal desilicification, alkali leaching and oxidation in metapelites of the Rosebel gold district in the Paleoproterozoic Marowijne Greenstone Belt, Suriname. In: Proceedings 12th Inter Guiana Geological Conference, Georgetown, Guyana, 115.

Naipal R, Kroonenberg S, Kriegsman LM, Mason PRD (2023) Oceanic-plateau ultramafic lavas and arc-related intrusives in two contrasting ultramafic complexes from the Paleoproterozoic Guiana Shield. Poster #154 Goldschmidt conference, Lyon, France

Ramlal S, Kroonenberg SB, Mason PRD, Kriegsman LM, O'Sullivan P (2019) Multiphase TTG intrusions in the Paleoproterozoic greenstone belt of Suriname and their role in gold mineralization in the Rosebel gold district. In: Proceedings 11th Inter Guiana Geological Conference, Paramaribo. Mededeling Geologisch Mijnbouwkundige Dienst Suriname 29:159–162

Reis NJ, Fraga LM, de Faria MSG, Almeida ME (2003) Geologia do Estado de Roraima, Brasil. Géologie de la France 2–3–4: 121–134.

Reis NJ, Dreher A, Fraga LM, Scandolara JE, Betiollo L. (2009) Serra Tepequém, um possível remanescente de uma caldeira vulcânica paleoproterozóica – Estudos preliminares. In: Simpósio de Geologia da Amazônia, XI, 175–195 aetro. 46, 169–188 (1974)ntr. Mineral. and Petrol. 46, 169–188

Reis NJ, Teixeira W, D'Agrella-Filho MS, Bettencourt JS, Ernst RE, Goulart LEA (2021) Large igneous provinces of the Amazonian Craton and their metallogenic potential in Proterozoic times. In: Srivastava RK, Ernst RE, Buchan KL, de Kock M (eds) Large Igneous Provinces and their plumbing systems. Geological Society, London, Special Publications 518:493–529

Rosa-Costa LT, Ricci PSF, Lafon J-M, Vasquez ML, Carvalho JMA, Klein EL, Macambira EMB (2003) Geology and geochronology of Archean and Paleoproterozoic domains of southwestern Amapá and northwestern Pará, Brazil, southeastern Guiana shield. Géologie de la France 2–4:101–120

Rosa-Costa LT, Lafon J-M, Delor C (2006) Zircon geochronology and Sm–Nd isotopic study: Further constraints for the Archean and Paleoproterozoic geodynamical evolution of the southeastern Guiana Shield, north of Amazonian Craton, Brazil. Gondwana Res 10:277–300

Rosa-Costa LT, Lafon J-M, Cocherie A, Delor C (2008) Electron microprobe U-Th-Pb monazite dating of the Transamazonian metamorphic overprint on Archean rocks from the Amapá Block, southeastern Guiana Shield, Northern Brazil. J S Am Earth Sci 26:445–462

Rosa-Costa LT, Monié P, Lafon J-M, Arnaud NO (2009) [40]Ar[39]Ar geochronology across Archean and Paleoproterozoic terranes from southeastern Guiana Shield (north of Amazonian Craton, Brazil): Evidence for contrasting cooling histories. J S Am Earth Sci 27:113–128

Sastrohardjo F, Vanderhaeghe O, Kriegsman L, Eglinger A, Kroonenberg S, Bardoux M (2022) Nature of the relationship between the Marowijne greenstone belt and the Gran Rio granite of the Rhyacian Transamazonian orogenic belt, Suriname: Significance of the Sara's Lust migmatite. In: Proceedings SAXI- XII Inter-Guiana Geological Conference 2022: Georgetown, Guyana, 128–134

Tassinari CCG, Munhá JMU, Teixeira W, Palacios T, Nutman A, Sosa SC, Santos AP, Calado BO (2004) The Imataca Complex, NW Amazonian Craton, Venezuela: crustal evolution and integration of geochronological and petrological cooling histories. Episodes 27:3–12

Tedeschi M, Hagemann S, Kemp AIS, Kirkland CL, Ireland TR (2020) Geochronological constrains on the timing of magmatism, deformation and mineralization at the Karouni orogenic gold deposit: Guyana. South America. Precambrian Research 337:105329

Tedeschi M, Perrouty S, Fredericks L, Bardoux M (2022) Preliminary lithostratigraphy of the Rhyacian Greenstone Belts of Northern Guyana and Suriname SAXI- XII Inter-Guiana Geological Conference 2022: Georgetown, Guyana, 144–145

Thébaud M, Tedeschi M (2022) Rhyacian crustal evolution of the Guyana Shield revealed through U-Pb and Lu-Hf analyses. In: Proceedings SAXI- XII Inter-Guiana Geological Conference, Georgetown, Guyana, 146–150

Vanderhaeghe O, Ledru P, Thiéblemont D, Egal E, Cocherie A, Tegyey M, Milési JP (1998) Contrasting mechanism of crustal growth. Geodynamic evolution of the Paleoproterozoic granite–greenstone belts of French Guiana. Precambr Res 92:165–193

Veenstra E (1983) Petrology and geochemistry of sheet Stonbroekoe, sheet 30, Suriname. Thesis, University of Amsterdam. Also. Mededelingen Geologisch Mijnbouwkundige Dienst Suriname 26:1–138

Velázquez GD, Béziat S, Salvi T, Tosiani DP (2011) First occurrence of Paleoproterozoic oceanic plateau in the Guiana Shield: The gold-bearing El Callao Formation, Venezuela. Precambr Res 186:181–192

Verhofstad J (1971) The geology of the Wilhelmina Mountains in Suriname, with special referene to the occurrence of Precambrain ash-flow tuffs. Thesis Univ. Amsterdam. Mededeling Geologisch Mijnbouwkundige Dienst Suriname 21:9–97

Voicu G, Bardoux M, Stevenson R (2001) Lithostratigraphy, geochronology and gold metallogeny in the northern Guiana Shield, South America: a review. Ore Geol Rev 18:211–236

Wijngaarde GW, Kroonenberg SB, Mason PRD, Kriegsman LM (2019) Petrography, geochemistry and age of the Armina Formation metaturbidites of the Coppename River, Suriname. In: Proceedings 11th Inter Guiana Geological Conference, Paramaribo. Mededeling Geologisch Mijnbouwkundige Dienst Suriname 29, 201–205

Chapter 10
The Guiana Shield: Reworked Older Basement or Juvenile Additions?

Abstract There is as yet no simple answer to the question posed at the beginning of this book: how much is reworked older basement, how much is juvenile? Recent Sm-Nd and zircon Lu-Hf model studies and the common presence of inherited Archean zircons suggest that it is more complicated than the simple continental growth model advocated in the past. New data are expected in the coming years as a result of international cooperation programmes. That some form of plate tectonics has operated is generally accepted, but whether it is similar to the present processes or forms a transition between Archean conditions and modern tectonics is also still an open question.

In the whole Amazonian Craton Archean rocks only crop out in three relatively small areas: the Imataca belt and the Amapá blocks in the Guiana Shield, and the Carajás block in the Guaporé Shield. Do these areas represent the last visible remains of a continuous Archean craton that was later reworked by Proterozoic processes? Or did the vast stretches of Paleoproterozoic rock units originate directly from differentiation from the mantle? Already Martín-Bellizzia (1972) and Mendoza (1973) differed in opinion about these questions. There are several arguments that have to be considered.

Most authors now accept that some form of plate tectonics must have operated in the Paleoproterozoic. The Trans-Amazonian Orogeny produced island arcs and ultimately continental collision over the whole 1500 km length of the Maronian greenstone belt in a very short time, mainly between 2.18 and 2.08 Ga. The sheer length of the belt and its internal parallellism are difficult to understand without assuming that there was some kind of continental cratonic hinterland. Most plate tectonic reconstructions, including Mendoza (1973), Maas (1979) and Delor et al. (2003) visualise the orogeny as the result of a collision between ancestral Amazonian or Guianian and West-African plates. There is also paleomagnetic evidence for this hypothesis (D'Agrello-Filho et al. 2016, Reis et al. 2021).

A second argument for a role for a larger pre-Trans-Amazonian Archean craton is the common presence of Archean detrital zircons in both igneous and supracrustal sequences in the greenstone belts and the high-grade belts in large parts of the Guiana Shield. In the Paleoproterozoic greenstone belt of Suriname Archean detrital zircons have been found in Paramaka metarhyolites, Armina metagreywacke and Rosebel arenites and conglomerates (Daoust 2016), in TTG bodies and in rhyolite ash layers intercalated into the Rosebel arenites (Daoust 2016; Ramlal et al. 2019), in late S-type potassic granites (Kromopawiro et al. 2019). Fraga et al. (2024) demonstrate that 103 out of 360 spots analysed on zircons from the Cauarane-Coeroeni Belt, the Rio Urubu Belt and the Orocaima belt in the central part of the Guiana Shield show an Archean provenance. The oldest zircons encountered in the shield give an age of 4219 ± 19 Ga, and have been found in felsic magmatic rocks of the Iwokrama Formation in Guyana corresponding to the Orocaima SLIP (Nadeau et al. 2013, 2022).

In the early days of isotope geochronology, low initial $^{87}Sr/^{86}Sr$ ratios around 0.700–0.705 in Rb-Sr whole rock isochrons were taken as evidence of a juvenile origin of the dated rocks, because such values were considered as similar to those of the mantle from which they originated. Higher ones were considered to indicate the involvement of older crust (e.g. Cordani et al. 1979, 1988; Tassinari 1981; Bosma et al. 1983; Gibbs and Barron 1983). At present the Rb–Sr method is largely abandoned, because of the mobility of the elements and hence the possibility of open system behaviour, and the low blocking temperatures in comparison with other systems.

This approach has now been superseded by Sm-Nd model studies (Gruau et al. 1985; Tassinari 1996; Cordani and Sato 1999, cf Chap. 6). The initial ratio $^{143}Nd/^{144}Nd$ is expressed as εNd (t), which indicates 10,000 times the deviation of the initial ratios from the CHUR (Chondrite Uniform Reservoir) evolution line (Dickin 2005). During crustal extraction from the mantle the Sm/Nd ratio of the mantle increases due to preferential uptake of the lighter Nd in crustal material, and therefore a completely juvenile magma should not plot on the CHUR evolution line, but on the depleted mantle line. The Sm-Nd model age T_{DM} of a series of rocks, i.e. the time of extraction of their magmas from the mantle, is found by connecting their εNd value plots and find the intersection with the depleted mantle line. In general, rocks with positive εNd values can be considered as largely juvenile, and with negative εNd as recrystallised from older crustal material. An early application of the method has been presented by Gruau et al. (1985) for the greenstone belt in French Guiana, suggesting juvenile sources for the (ultra)mafic and intermediate volcanics in the greenstone belt (Fig. 10.1).

Cordani and Sato (1999) applied the procedure for granite samples from the Amazonian Craton in Fig. 6.6. As stated above, most of their granites plot around the CHUR line, implying largely juvenile provenance. Velázquez et al. (2011) consider most of the mafic volcanics in the Venezuelan greenstone belt as juvenile on the basis of mainly positive εNd values. Klaver et al. (2015) showed that the intrusive 1.98 Ga Orocaima charnockites in the Bakhuis Granulite Belt have no Archean ancestry, showing CHUR εNd values and 2.22–2.34 Ga T_{DM} model ages, and De Roever et al. (2015) and Barbosa et al. (2021) reach similar conclusions on other Orocaima felsic volcanics ands hypabyssal granites.

A still more sophisticated approach has been developed on the basis of the Lu-Hf system. In this case it is not necessary to analyse the whole granitic rock, because almost all Hf is contained in the mineral zircon. Zircon separates are usually already

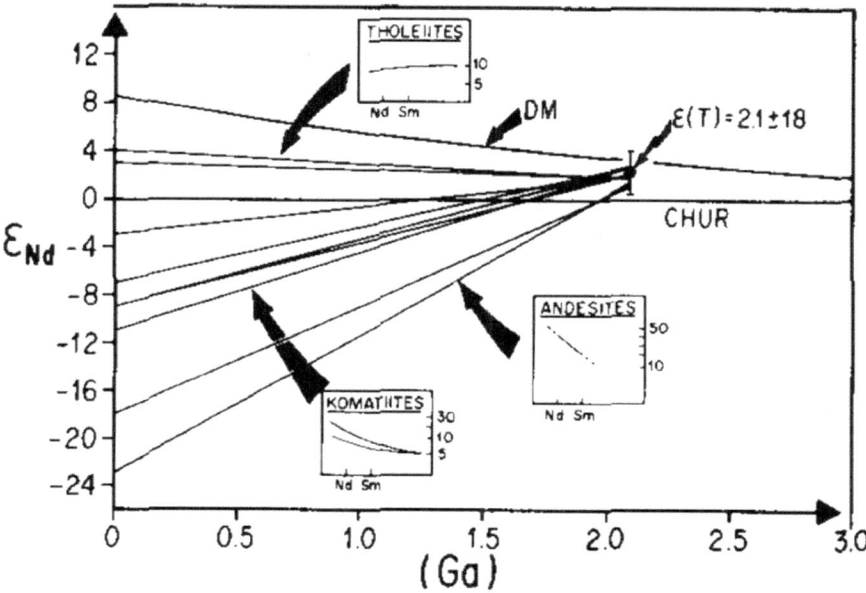

Fig. 10.1   Nd isotopic evolution diagram showing eNd variation as a function of time for the studied samples. Schematic REE abundance patterns for each compositional type are shown as insets linked to growth lines by solid arrows. Legend: CHUR-chondritic uniform reservoir. DM-Depleted mantle evolution line (Gruau et al. 1985)

available from earlier U-Pb dating. In the same way as for the Sm-Nd system the initial ratio for $^{176}Hf/^{177}Hf$ can be expressed as εHf, and T_{DM} model ages can be calculated accordingly.

A combined Sr-Nd-Hf study on the Amapá block and surroundings by Lafon et al. (2019) and Milhomem Neto and Lafon (2019) showed at least two Archean episodes of crust generation around 4.0 Ga and 3.1–3.0 Ga. In the Rhyacian Lourenço and Carecuru greenstone terrains magmatic rocks are derived from mixed Rhyacian juvenile sources with Archean crustal components. The older Orosirian magmatic province of 2.0–1.97 Ga, geochemically testifying of an orogenic tectonic setting, (coinciding with the Orocaima SLIP) show isotopic evidence of melting from mantle-derived magmas with participation of Rhyacian sialic crust. In the younger magmatic province of 1.90–1.87 Ga, coinciding with the Uatumã SLIP and the famous 'Central Amazonian Province', Archean model ages and inherited zircons are lacking. Their data are incorporated in Fig. 10.3.

An interesting series of Hf isotope data has been presented by Ibáñez-Mejía (2014). Along a stretch of 1000 km along the Orinoco river, the westernmost border of the Guiana Shield, he sampled zircons from all major blocks of the shield, including the Imataca Belt, the Pastora part of the greenstone belt in Venezuela, Cuchivero granites of the Orocaima Belt, gneisses from the Rio Negro belt and the young Parguaza rapakivi (Fig. 10.2). He postulates a Guyana Primordial Crust of around

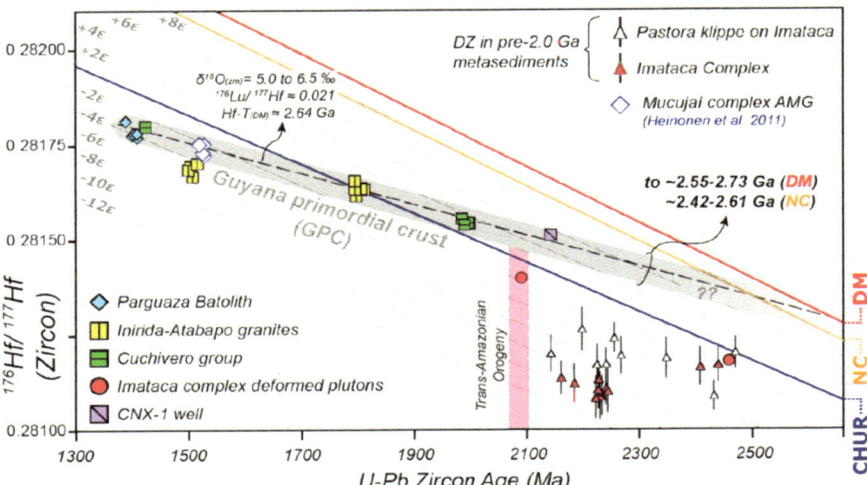

Fig. 10.2 Initial ^{176}Hf/^{177}Hf compositions of analyzed zircons at their U–Pb age of crystallization. The light gray area represents the Guiana Primordial Crust of ~2.6 Ga (GPC) (Ibáñez-Mejía 2014)

2.6 Ga, which corresponds with the frequency with which detrital zircons of that age are found in many areas.

Also in the Río Negro Block, there is regular incorporation of older Rhyacian crustal material into the granitoid rocks, as shown by Ibáñez-Mejía and Cordani (2020) (Chapter 8).

Lu-Hf data acquired during the SAXI project from greenstone belt intrusives in Suriname and Guyana show mainly superchondritic εHf values comprised between − 6.15 and + 4.89 pointing toward juvenile crust extraction from a depleted mantle source (T_{DMc} = 2.2 to 2.4 Ga) with little contribution from Archean crustal material (Thébaud and Tedeschi 2022).

To finalise this chapter and this book, the paper of Thébaud and Tedeschi (2022) clearly expresses to where all this boils down: 'Our initial study of the crustal evolution of the Guiana shield suggests that the Paleoproterozoic continental lithosphere developed through a major episode of juvenile mantle extraction at ca. 2160 Ma in the vicinity of pre-existing Archean crustal blocks (Amapá and Imataca complexes). Ongoing plutonism that spans nearly ~200 Ma lead to further progressive reworking of the juvenile Paleoproterozoic crust in a convergent tectonic setting. Such evolution mimics that recorded in the West African Craton which preserves Rhyacian granite-greenstone belts of similar age.'

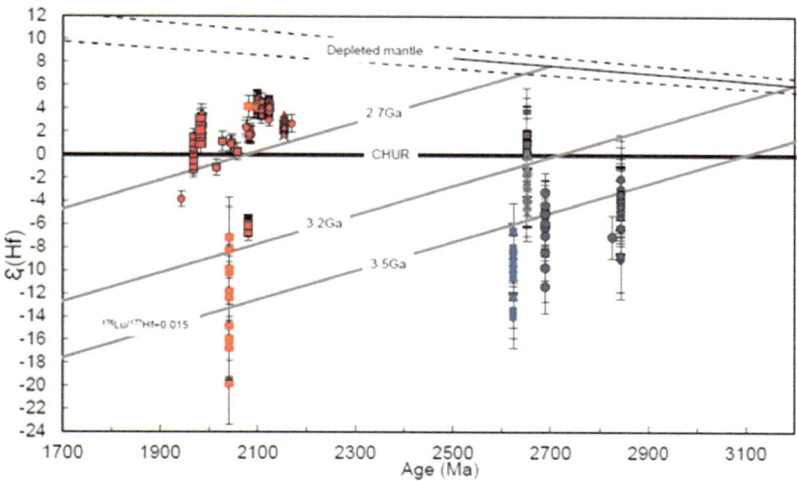

Figure 3: e(Hf)vs age plot for new (red) and published (grey) data. Published data were sourced from Neto and Lafon, 2019.

Fig. 10.3 εHf age plot for new SAXI data from Guyana (red) and old data from Amapá (grey) published by Milhomem Neto and Lafon (2019) (Thébaud and Tedeschi 2022)

References

Barbosa NA, Fuck RA, Souza VS, Dantas EL, Tavares Júnior SS (2021) Evidence of a Palaeoproterozoic SLIP, northern Amazonian Craton, Brazil. J S Am Earth Sci 111:103453

Bosma W, Kroonenberg SB, Maas K, de Roever EWF (1983) Igneous and metamorphic complexes of the Guiana shield in Suriname. Geol Mijnbouw 62:241–254

Cordani UG, Tassinari CCG, Teixeira W, Basei MAS, Kawashita K (1979) Evolução tectônica da Amazonia com base nos danos geocronológicos. 2ndo Congreso Geológico Chileno, 139–148

Cordani UG, Sato K (1999) Crustal evolution of the South American Platform, based on Nd isotopic systematics on granitoid rocks. Episodes 1999:167–173

D'Agrella-Filho MS, Bispo-Santos F, Trindade RIF, Antonio PYJ (2016) Paleomagnetism of the Amazonian Craton and its role in paleocontinents. Braz J Geol 46:275–299. https://doi.org/10.1590/2317-4889201620160055

Daoust C (2016) Caractérisation stratigraphique, structurale et géochimique du District minéralisé de Rosebel (Suriname) dans le Cadre de l'évolution géodynamique du Bouclier Guyanais, PhD thesis, Université du Québec á Montréal (Montréal): 330

Delor C, de Roever EWF, Lafon JM, Lahondère D, Rossi Ph, Cocherie A, Guerrot C, Potrel A (2003b) The Bakhuis ultra-high temperature granulite belt Suriname : II implications for late Trans-Amazonian crustal stretching in a revised Guiana Shield framework. Géologie de la France 2–4:207–230

De Roever EWF, Lafon J-M, Delor C, Cocherie A, Guerrot C (2015) Orosirian magmatism and metamorphism in Suriname: new geochronological constraints. Contribuições á Geologia Da Amazônia 9:359–372

Dickin AP (2005) Radiogenic isotope geology. Cambridge University Press, 2nd edn, pp 492

Fraga LM, Cordani UG, Dreher AM, Sato K, Reis NJ, Nadeau S, De Roever E, Kroonenberg S, Camara Maurer VC (2024) Early Orosirian belts of the central Guiana Shield, northern

Amazonian Craton: U-Pb geochronology and tectonic implications Precambrian Research 407:107362

Gruau G, Martin H, Leveque B, Capdevila R (1985) Rb-Sr And Sm-Nd Geochronology of Lower Proterozoic granite–greenstone terrains In French Guiana, South America. Precambr Res 30:63–80

Ibánez-Mejía M, Urbani F, Gehrels GE, Valley J, Pullen A, Ducea MN, Pepper M, Ruiz, J (2014) Episodic Archean generation of the Guyana Shield crust followed by one billion years of apparent no-growth. In: Timing and rates of Precambrian crustal genesis and deformation in northern South America. PhD thesis University of Arizona, pp 161–234

Ibáñez-Mejía M, Cordani, UG (2020) U–Pb Geochronology and Hf–Nd–O Isotope Geochemistry of the Paleo–to Mesoproterozoic Basement in the Westernmost Guiana Shield. In: Gómez J, Mateus–Zabala D (eds) The Geology of Colombia, Volume 1 Proterozoic–Paleozoic. Servicio Geológico Colombiano, Publicaciones Geológicas Especiales 35:65–90. https://doi.org/10.32685/pub.esp.35.2019.04

Klaver M, De Roever EWF, Nanne JAM, Mason PRD, Davies GR (2015) Charnockites and UHT metamorphism in the Bakhuis Granulite Belt, western Suriname: Evidence for two separate UHT events. Precambr Res 262:1–19

Kromopawiro S, Kroonenberg SB, Kriegsman LM, Mason PRD (2019) 2.12–2.08 Ga Late- to post-collisional peraluminous granitoid magmatism in the Marowijne Greenstone Belt of Suriname. In: Proceedings 11th Inter Guiana Geological Conference, Paramaribo. Mededeling Geologisch Mijnbouwkundige Dienst Suriname 29:105–109

Lafon JM, Rosa-Costa L, Milhomem Neto JM (2019) Sr-Nd-Hf isotopic tracing of Archean continental crust in the Brazilian part of the Southeastern Guyana Shield: A review. In: Proceedings 11th Interguiana Geological Conference 2019: Paramaribo, Suriname. Mededeling Geologisch Mijnbouwkundige Dienst Suriname 29:121–124

Maas K (1979) Nota betreffende een overzicht, alsmede een tentatieve interpretatie van het Precambrium van Suriname. Geologisch Mijnbouwkundige Dienst Suriname, internal report, pp 23

Martín-Bellizzia C (1972) Paleotectónica del Escudo de Guayana. Memoria de la 9na conferencie geológica Inter-Guayanas. Boletín de Geología (Caracas). Publicación Especial No. 6:251–304

Mendoza V (1973) Evolución tectónica del Escudo de Guayana. Segundo Congreso Latinoameri-cano de Geologia, Boletín Geológico Caracas Publicación Especial. 7, III: 2237–2270

Milhomem Neto JM, Lafon JM (2019) Zircon U-Pb and Lu-Hf isotope constraints on Archean crustal evolution in southeastern Guyana Shield. Geosci Front 10:1477–1506

Nadeau S, Chen W, Reece J, Lachhman D, Ault R, Faraco MTL, Fraga LM, Reis NJ, Betiollo LM (2013) Guyana: The lost hadean crust of South America? Braz J Geol 43:601–606

Nadeau S, Fredericks L, Reece J (2022) Update on the Guyana-Brazil project and more recent zircon age results of alkaline intrusions in Southern Guyana: Makarapan Mountain and Muri Mountain Alkaline Complex. SAXI-XII Inter-Guiana Geological Conference 2022: Georgetown, Guyana, pp 101–104

Ramlal S, Kroonenberg SB, Mason PRD, Kriegsman LM, O'Sullivan P (2019) Multiphase TTG intrusions in the Paleoproterozoic greenstone belt of Suriname and their role in gold mineral-ization in the Rosebel gold district. In: Proceedings 11th Inter Guiana Geological Conference, Paramaribo. Mededeling Geologisch Mijnbouwkundige Dienst Suriname 29:159–162

Reis NJ, Teixeira W, D'Agrella-Filho MS, Bettencourt JS, Ernst RE, Goulart LEA (2021) Large igneous provinces of the Amazonian Craton and their metallogenic potential in Proterozoic times. In: Srivastava RK et al., (eds). Large Igneous Provinces and their plumbing systems. Geological Society, London, Special Publications, 518, 493–529

Tassinari CCG (1981) Evolução geotectônica da Província rio Negro-Juruena na região amazônica. Instituto de Geociências, Universidade de São Paulo, Dissertação de mestrado, p 99

Tassinari CCG (1996) O mapa geocronológico do Cráton Amazônico no Brasil: revisão dos dados isotópicos. Universidade de São Paulo, Tesis de livre-docência, p 257

Thébaud M, Tedeschi M (2022) Rhyacian crustal evolution of the Guyana Shield revealed through U-Pb and Lu-Hf analyses. In: Proceedings SAXI-XII Inter-Guiana Geological Conference, Georgetown, Guyana, pp 146–150

Velázquez GD, Béziat S, Salvi T, Tosiani DP (2011) First occurrence of Paleoproterozoic oceanic plateau in the Guiana Shield: The gold-bearing El Callao Formation, Venezuela. Precambr Res 186:181–192